Focused Beam Methods

Measuring Microwave Materials in Free Space

JOHN W. SCHULTZ
Atlanta GA, USA

ACKNOWLEDGEMENTS

Special thanks go to my colleague, Stephen Blalock, who provided many suggestions and edits to this manuscript. Thanks also go to the many researchers at Georgia Tech whose research efforts have contributed to the material in this book. In particular, significant contributions came from Dr. Jim Maloney, Mr. Ed Hopkins, Dr. Rick Moore, Professor Glenn Smith, and Dr. David Reid. Finally, much of this work would not have been possible without the contributions from the many graduate students and undergraduate co-op students who have worked in our GTRI facilities.

CONTENTS

PREFACE

Determining the intrinsic microwave properties of materials is important for a variety of applications ranging from antenna and electronic circuit design to remote sensing to electromagnetic interference mitigation. A number of methods exist for characterizing intrinsic properties of materials at microwave frequencies, including transmission lines, resonant cavities, and impedance analysis. The use of free-space measurement methods has become commonplace among microwave material characterization laboratories due to its ease of use and reasonable accuracy. While some free-space facilities exist that can characterize down to 500 MHz, the method is most useful for characterizing materials from 2 GHz through millimeter waves. This book is designed to acquaint engineers and scientists with the theory and practice of using microwave focused beam systems for free-space characterization of materials. It assumes that the reader already has an undergraduate level knowledge of calculus and electrical engineering principals.

Free-space methods have a distinct advantage over other guided-wave transmission line methods because there is no interface between the material specimen and a guide wall. For example, coaxial airline and rectangular waveguide methods must completely span the space within the waveguide with the material. If any air-gap exists between the specimen and the walls of the waveguide, then microwave energy can preferentially travel through this gap rather than through the material specimen, resulting in sometimes exceedingly large systematic errors. In focused beam free-space systems in particular, the energy is confined to an area within a larger specimen so that no air-gap or edge effects exist. Another advantage of free-space systems is in their ability to handle inhomogeneous composites (such has honeycomb core) where a small waveguide or coaxial airline fixture does not measure a large enough region of material to get a statistically significant 'average' of the macroscopic properties.

While free-space methods, including focused beams, are among the most popular for determining dielectric and magnetic properties of materials at microwave frequencies, they also have some disadvantages. For example, relatively large specimen sizes are needed, which can be troublesome when large specimens are expensive or impractical to manufacture. Free space methods, and transmission line methods in general, also have difficulty obtaining accurate loss properties of low-loss specimens. Only resonant methods will typically have the fidelity and sensitivity to obtain accurate loss factors at low levels. In a nutshell, no single measurement device can address all the desired requirements for electromagnetic material characterization, and the free space methodologies are no exception to this rule. The numerous applications for electromagnetic materials, combined with the wide range of material configurations and characteristics provide a large number of variables that must be addressed by any material measurement laboratory. Frequency range, specimen size/shape, temperature, anisotropy, and inhomogeneity are some of the many variables that influence a given measurement scenario, and they determine which of the many measurement methodologies are most appropriate for a given specimen.

Because of the variety of potential variables associated with a material measurement, an effective measurement laboratory will utilize a number of different methods to satisfy the data needs of their customers. Each electromagnetic measurement fixture can be thought-of as a different tool in the measurement lab's toolbox, and having a variety of tools provides the flexibility to address the many applications and material configurations that exist. The objective of this book is to describe one of those tools: the free-space focused beam method, and to provide the necessary foundation for using this tool to accurately characterize microwave materials. It is intended to be a practical resource that contains the necessary components for understanding and implementing free-space measurement methods.

Specifically, this book reviews the use of free-space characterization methods for microwave materials, with emphasis on focused beam methodologies. This first section provides a very brief overview of dielectric and magnetic material properties and highlights the history of free-space measurement development. The following chapter then describes the theory of Gaussian optics and how focused beam systems are typically implemented. This chapter also reviews calibration procedures and time-domain post-processing in free space measurements.

The latter is usually necessary for obtaining reasonable measurement accuracy because of discontinuities introduced by components such as the feed antenna and lens. Chapter three outlines many of the common inversion methods used to convert the calibrated scatter data into intrinsic material parameters (e.g. permeability -- μ and permittivity -- ε). Some of these inversion methods are then illustrated in chapter four, which shows example measurement data for both dielectric and magnetic materials. Understanding the accuracy of a fixture is critical for quality measurements in any situation, so chapter five discusses some of the sources for errors in focused beam measurement systems. Chapter six provides an overview of the use of a focused beam to characterize diffuse scatter from materials such as composites. An engineered composite consists of two or more constituent materials, and many of these composites can scatter not just specular energy, but diffuse energy as well, due to intrinsic inhomogeneity. This chapter demonstrates the use of a focused beam system to measure this "fuzz-ball" scatter both for composites as well as for materials with discontinuities (e.g. edges or gaps). Finally, the last chapter describes some advanced methods that can be applied to or with free space measurements.

While free-space measurement methods have been under development since the 1940s, there are still new techniques to be developed and new ways to apply the method to materials. With advances in computational tools and materials technology, the measurement of electromagnetic materials remains an active area of research and development. This book provides an overview of the most important ways that free-space systems are used, and provides a foundation for continued advancement of the state of the art in electromagnetic material measurements.

J.W. Schultz
October 2012

1. INTRODUCTION

DIELECTRIC AND MAGNETIC PROPERTIES OF MATERIALS

Before delving into the measurement of intrinsic electromagnetic materials properties, a certain level of insight about the physical mechanisms behind these intrinsic properties is required. The origin of important material parameters and their influence on the macroscopic characteristics provide guidance for how best to apply free-space measurement and analysis to a given material system. This section reviews some of the fundamental aspects of materials and their influence on both dielectric and magnetic properties.

For any material, whether it is a polymer, ceramic, or metal, an applied electric field induces an electric polarization within the material. Usually the magnitude of this polarization is linearly proportional to the applied field. The proportionality constant in this case is called the permittivity and is designated by the symbol, ε. A fundamental constant of nature is the permittivity of vacuum, $\varepsilon_0 = 8.854 \times 10^{-12}$ F/m. Usually the permittivity is expressed as the relative permittivity, which is the ratio of the material permittivity to the permittivity of a vacuum, $\varepsilon_r = \varepsilon/\varepsilon_0$. This can often be a source of confusion since it is common to drop the subscript 'r' from the symbol for relative permittivity. It is usually up to the reader to infer whether 'ε' means permittivity or relative permittivity based on its context. To add to the confusion, the relative permittivity is sometimes also called simply 'permittivity' or the 'dielectric constant'. The convention of this book is to leave off the subscript 'r' and the dielectric permittivities (and magnetic permeabilities) are assumed to be relative unless otherwise designated.

In a time varying or oscillating electric field, the permittivity is best represented by a complex number, $\varepsilon = \varepsilon' - i\varepsilon''$. In this notation, ε' is the real part of the permittivity, and is often called 'permittivity' for short. ε'' is the imaginary part of the permittivity, and is usually called the dielectric 'loss factor'. The loss factor is usually associated with energy absorption by the material. With the definition above, ε'' should always be a positive number since energy conservation dictates that a passive material cannot exhibit gain. In some cases, the complex permittivity is defined with a '+' instead of a '-' (i.e. $\varepsilon = \varepsilon' + i\varepsilon''$), in which case the ε'' will be a negative number. In this book the '-' convention is used to define the complex permittivity so that the loss factor should not be negative.

Another quantity that is associated with the energy absorption by a material is the 'loss tangent', defined by $\tan \delta = \varepsilon''/\varepsilon'$. This loss tangent is another way to express how a material absorbs energy and is simply the tangent of the angle defined by the real and imaginary permittivity in the complex plane. Because it effectively normalizes the loss factor by the real part of the permittivity, it can be a convenient way to compare the dielectric loss of materials that have differing real permittivities. Yet another definition is useful in the case where conduction processes are occurring, where it is possible to define an 'apparent' conductivity. This quantity, σ, is usually calculated from the dielectric loss factor by, $\sigma = \omega\varepsilon'' = \omega\varepsilon_0\varepsilon''_r$, where ω is the angular frequency, $\omega = 2\pi f$.

When analysing the dielectric data of a given material, there are two dominant physical phenomena that can contribute to the observed complex permittivity: dipole reorientation and conduction. In simple terms, dipoles are manifested by charge asymmetries that exist in crystals, molecules or molecular fragments. A molecular fragment that has at least two different atoms can have an intrinsic dipole moment because different atoms have different affinities for electric charge. Example dipoles are illustrated in Figure 1, which shows a hypothetical polymer fragment on the left and an array of ionically bonded atoms (e.g. a ceramic) on the right. Dipole moments exist between atoms with opposite charge, and these dipoles are vectors that describe the charge distribution in units of [charge x displacement]. Polymer chains are usually made up of thousands of bonds that can rotate in response to external stimuli. Thus in an externally applied electric field, the dipole fragments will realign themselves to partially cancel the effects of the electric field. In a ceramic material, the charge centers will simply displace from their equilibrium position when an electric field is present. In essence, an applied electric field causes the electron cloud to shift relative to the nuclei thus creating a spatial charge imbalance.

The time that it takes for the dipole to realign itself varies according to the properties of the material and external conditions such as temperature and pressure. Thus a material's dielectric response can also be characterized by the relaxation time (or more precisely, the distribution of relaxation times) of the intrinsic dipoles. In terms of frequency, shorter relaxation times correspond to higher frequencies while longer relaxation times correspond to lower frequency behaviours.

The second way a material responds to an electric field is conduction, which involves the physical translation of charged species. Charged species can be either ions or electrons. When an electric field is applied, opposites attract, so a positively charged species will be attracted to the negative potential, while the negatively charged species will be attracted to the positive potential. More precisely, an applied electric field perturbs the Brownian motion of ions within a material so that they tend to drift towards oppositely charged electrodes depending on their charge.

polymeric material   ceramic material

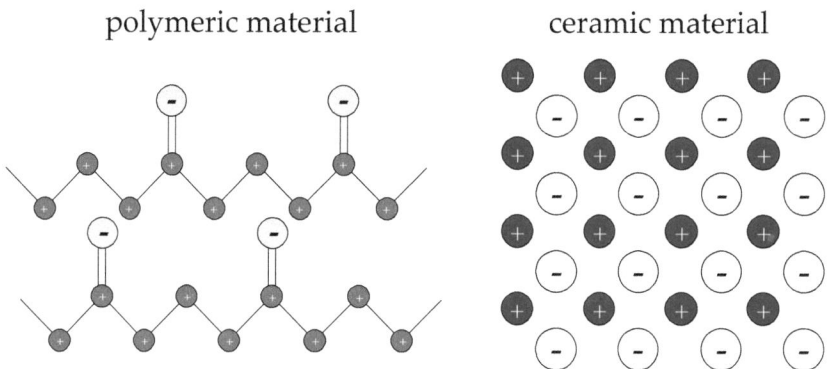

Figure 1   Schematic representation of charge distribution, which leads to dipoles within different types of materials.

Similar to dipole reorientation, ion (as well as electron) conduction is also affected by various chemical and environmental variables. As the charged species travel towards a positive potential, they are slowed by their surroundings. The slower the ions travel, the more resistant the material. The faster the ions travel, the more conductive the material. Electron conduction happens when there are electrons that are not bound to nuclei. These unbound electrons are prevalent in graphitic materials and metals. Like ionic conduction, electron conduction can be affected by imperfections in the crystal lattice, temperature, or pressure. The parameter that quantifies how well electrons and ions can travel is called conductivity.

Magnetic properties in a material are a response to an externally applied magnetic field. Since microwave radiation includes oscillating electric and magnetic fields, magnetic materials can strongly interact with incident microwave energy. Magnetic permeability is denoted by the symbol, μ, which is the proportionality factor that relates the magnetic flux density to the magnetic field, and which depends on intrinsic material properties such as magnetic moment, domain magnetization, etc. A fundamental constant is the permeability of vacuum, $\mu_0 = 4\pi \times 10^{-7}$ H/m. Like permittivity, the magnetic permeability is usually expressed as the relative permeability, which is the ratio of the material permeability to the permeability of a vacuum, $\mu_r = \mu/\mu_0$. This can also be a source of confusion since it is common to drop the subscript 'r' from the symbol for relative permeability. It is usually up to the reader to infer whether 'μ' means absolute or relative permeability based on its context.

In a time varying or oscillating electric field, the permeability is best represented by a complex number, $\mu = \mu' - i\mu''$, where μ' is the real part of the permeability and μ'' is the imaginary part. Analogous to permittivity, μ'' is associated with energy absorption by the material interacting with the magnetic field and is called the magnetic loss factor. Also note that the sign convention used for permeability is the same at that for permittivity, and μ'' is always a positive number in this book. A magnetic loss tangent can also be defined as an alternate way to compare the loss associated with different magnetic materials, $\tan \delta_m = \mu''/\mu'$.

Most natural materials have little or no magnetic response such that their absolute magnetic permeability is equal to the permeability of free space. Magnetic moments within a material are primarily associated with electrons, and specifically with a quantum mechanical property of an electron called 'spin', which creates a small magnetic dipole moment associated with each electron. Since electrons share orbital shells when they are part of an atom or molecule, they tend to pair up with other electrons that have their spin oriented in the opposite direction. Thus the pairing of these electrons cancels out most of the magnetic dipole moment that could otherwise exist in a molecule. However in some substances, there can exist unpaired electrons within an orbital shell, and the net magnetic moment of these unpaired electrons, then leads to nontrivial magnetic permeability. This type of magnetic response is also known as ferromagnetism. Iron, iron alloys, and iron oxide ceramics are all examples of ferromagnetic materials that are microwave-frequency relevant.

DISPERSION AND ANISOTROPY

Measurement of the dielectric and magnetic properties of a material specimen can be complicated in and of itself. However, there are additional complications that occur, particularly in engineered materials, which make the job of determining intrinsic properties even more difficult. These additional complications include dispersion and anisotropy. Dispersion is the property of a material to have a frequency dependent dielectric permittivity and/or magnetic permeability. A notional dispersion curve for the dielectric permittivity is shown in Figure 2. The real permittivity, ε' is shown to undergo a step decrease as frequency increases while the imaginary permittivity, ε'' shows a peak corresponding to the maximum slope in ε'.

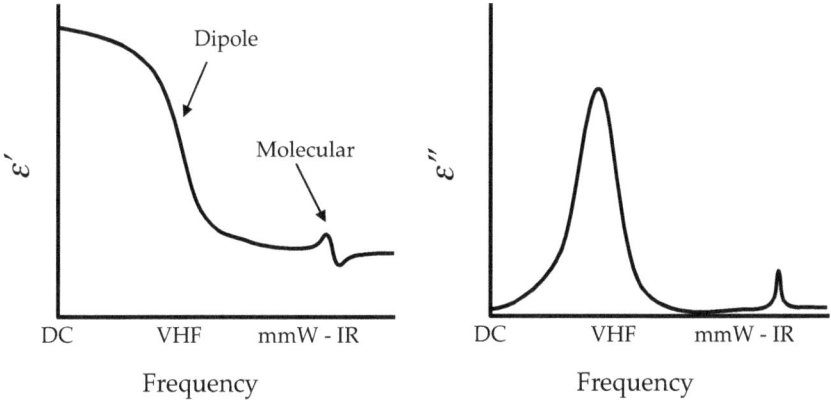

Figure 2   **Notional real and imaginary relative permittivity for a dispersive material.**

These changes are sometimes called 'relaxations', and are common in the dielectric permittivity and magnetic permeability of many electromagnetic materials. In fact, there are a number of analytical models that are commonly used to describe these relaxations, including the Debye and Lorentz models for dielectric and magnetic relaxations, and the Drude model for metals and other conductive materials [1,2,3]. These relaxation theories are classical models based on dipole displacement or conductor transport in a medium. They are summarized in the table below. Fitting measured data and comparing the fitted parameters is a more concise means for comparing different measurements. It also provides a convenient mechanism for defining dispersive models in computational electromagnetics codes used in design and prediction. Note that the present author uses the forms provided in the table, but other variations of these equations also exist.

For the expressions in Table 1, ε_U and ε_R are the high and low frequency limits (unrelaxed and relaxed) of the permittivity, τ is the characteristic relaxation time, ω_0 is a characteristic relaxation frequency, ω_p is another characteristic relaxation frequency called the "plasma frequency", δ is a damping parameter, and ω is the frequency in rad/s. The parameter, τ, is sometimes also expressed as a characteristic relaxation frequency, i.e. $\tau = 1/\omega_0$. Note that while these equations are specified for permittivity, they can also be applied to magnetic permeability. Typically the Lorentz dispersion model is most applicable to magnetic materials and can be used by replacing the dielectric permittivity parameters by corresponding magnetic permeability parameters.

Table 1   **Summary of classical dispersion theories.**

Relaxation Model	Expression
Debye	$\varepsilon = \varepsilon_U + \dfrac{\varepsilon_R - \varepsilon_U}{1 + i\omega\tau}$
Lorentz	$\varepsilon = \varepsilon_U + \varepsilon_R \dfrac{\omega_0^2}{\omega_0^2 - \omega^2 + 2i\omega\delta}$
Drude	$\varepsilon = \varepsilon_U - \dfrac{\omega_p^2}{\omega^2 - i\omega\delta}$

The models in Table 1 are ideal, and most materials do not exactly fit them. Often the relaxation phenomena in real materials occur over a wider bandwidth than these models predict, and additional empirical parameters are incorporated to improve the fit. For example, Cole and Cole generalized the Debye model by including an empirical exponent, α, [4]

$$\varepsilon = \varepsilon_U + \frac{\varepsilon_R - \varepsilon_U}{1 + (i\omega\tau)^{1-\alpha}} \qquad (1.1)$$

This exponent enables the relaxation to be extended over a broader range of frequencies. Figure 3 demonstrates the difference between the Debye function and the Cole-Cole variant by fitting these models to measured data for a graphite filled polyimide material known as Black Kapton®. The measured data are shown as a solid blue line in these plots, and the model fit data were calculated with a standard computational function minimization method. These data show the necessity for modifying the idealized dispersion models when trying to fit them to real data.

Going further Havriliak and Negami [5] defined an even more general variant to the Debye model,

$$\varepsilon = \varepsilon_U + \frac{\varepsilon_R - \varepsilon_U}{[1 + (i\omega\tau)^{\alpha}]^{\beta}} \qquad (1.2)$$

What Havriliak and Negami added, were the two empirical parameters, α and β, which allow for both flatness and asymmetrical skew in the relaxation behavior. Numerous researchers have proposed other empirical variations on these equations as well [6,7,8,9,10].

The data shown in Figure 3 are relatively low loss with an imaginary permittivity much smaller than the real part. This is characteristic of a material with bound charges, where dipole polarizability is the primary mechanism for dispersion. In some materials however, there may be un-bound electrons that conduct, which can lead to a dramatically different dispersion characteristic. For example Figure 4 shows measured data from a carbon-loaded foam, where the carbon loading is sufficient to create long-range conduction pathways for unbound electrons. In a material such as this, the imaginary permittivity can be significantly greater than the real permittivity, and more distinctly, the imaginary permittivity rises dramatically as measurement frequency is decreased.

In such a conductive material, it may make more sense to model the conductive dispersion with a conduction term.

$$\varepsilon = \varepsilon_U + \frac{\varepsilon_R - \varepsilon_U}{1 + i\omega\tau} + i\frac{\sigma}{\omega\varepsilon_0} \qquad (1.3)$$

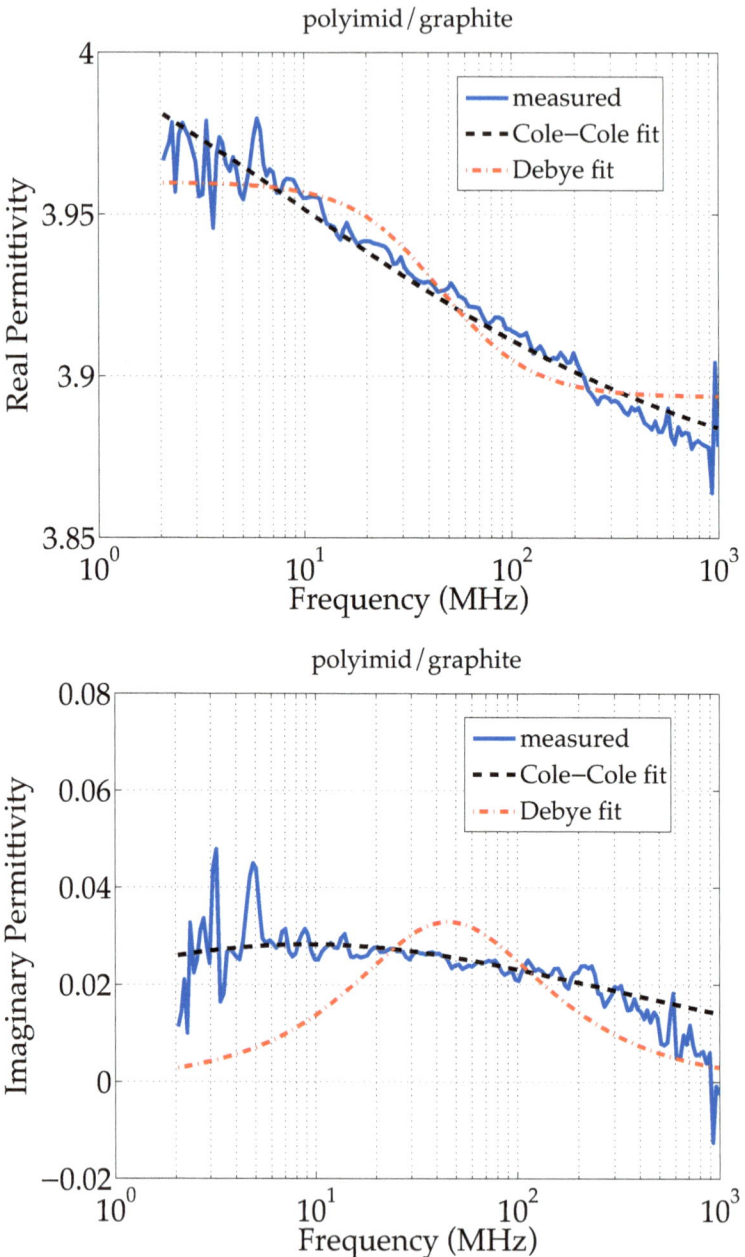

Figure 3   Comparison of model fits for the Debye and Cole-Cole
dispersion expressions. The Cole-Cole model fits the
measured data over a broader frequency range.

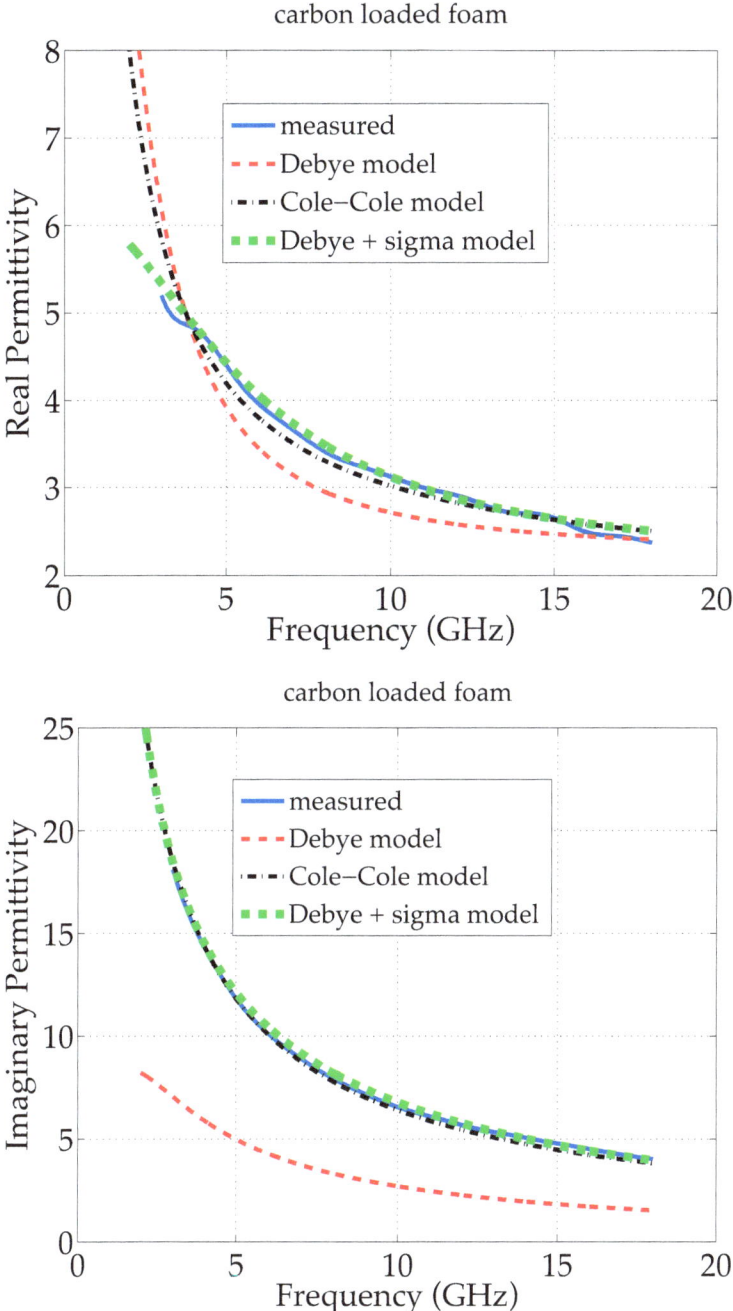

Figure 4   Comparison of measured data to three different
             dispersion models. The best fit in this case is with a
             Debye model that also has a conductivity term added.

where σ is the 'DC' conductivity and ε_0 is the permittivity of free space. In addition to the measured real and imaginary permittivity of a carbon loaded foam, Figure 4 shows several model fits to the data, including a simple Debye model, the Cole-Cole model, and the Debye plus conductivity term. The simple Debye model does not fit the data at all, which is not surprising since the Debye was originally derived as a model for materials with bound charges. The Cole-Cole function is significantly better, however it does not fit as well as the Debye plus conductivity term.

In the case of magnetic data, the Lorentz model can fit some data reasonably well, but many materials have too broad of a relaxation and require a more flexible dispersion model. A function that this author has used with some success to describe the frequency dispersive behavior of some magnetic composites is a double-Lorentz function. This double-Lorentz simply adds a second term to the usual Lorentz,

$$\mu = \mu_U + \frac{\omega_1^2(\mu_R - \mu_{Int})}{\omega_1^2 - \omega^2 + 2i\omega\delta_1} + \frac{\omega_1^2(\mu_{Int} - \mu_R)}{\omega_2^2 - \omega^2 + 2i\omega\delta_2} \tag{1.4}$$

where μ_U μ_{Int}, and μ_R are the unrelaxed (high frequency), intermediate, and relaxed (low frequency) permeabilities, ω_1 and ω_2 are the relaxation frequencies for the first and second terms, and δ_1 and δ_2 are the damping factors for the first and second terms in the above equation.

Understanding these dispersion models can provide insight into the measurement data acquired from a free-space measurement system. For example, dielectric relaxations that occur in the microwave range are usually slowly varying or follow a Debye-like frequency dependence. As such, the real part of the dielectric permittivity, ε', almost always decreases as frequency increases. Thus if a specimen measurement shows an increasing ε' with frequency, then the measurement data is likely suspect. On the other hand, the Lorentz model that many magnetic materials follow does allow for a rising real permeability with frequency over a part of the relaxation band. Another use for dispersion models is in extrapolation or gap filling. For example, if measurement equipment is only available in certain frequency bands, fitting the data to an appropriate dispersion model can provide an estimate of the material properties outside those measured frequencies or in gaps between measurement frequencies.

Another complicating factor in electromagnetic materials measurements is anisotropy. Anisotropy is the capability of some materials to have directionally dependent intrinsic properties. For many engineering composites, a single value of complex permittivity or permeability is insufficient as the permittivity depends on orientation of constituents within the material. For example, fiber reinforced composites with uniaxially oriented fibers can have a permittivity along the fiber direction that is markedly different than the permittivity orthogonal to the fiber direction. Particulate-filled materials may have in-plane properties that are different than out-of-plane properties when those particulates are shaped and aligned. Specifically in flake-filled materials, where the flakes are oriented "in-plane" the properties parallel to the plane of the flakes will be significantly different than the properties orthogonal to the flakes.

A third example, honeycomb core, has effective properties that are different in all three principal directions. Core consists of a tube-like geometry as shown in Figure 5. In this geometry there is a Length, Width, and Thickness direction, each having different values of permittivity corresponding to the geometrical differences in each direction. The thickness direction follows along the tubes, the width direction crosses some of the flats of the hexagonal tubes, and the length direction crosses the tubes but is parallel to some of the flats of the tubes. Thus an electric field oriented in any of these three directions will experience significant differences.

The prevalence of these anisotropic engineered materials in various electromagnetic applications means we must consider anisotropic materials where the dielectric permittivity and magnetic permeability are represented with three-by-three tensors,

$$\bar{\bar{\mu}} = \begin{bmatrix} \mu_{11} & \mu_{12} & \mu_{13} \\ \mu_{21} & \mu_{22} & \mu_{23} \\ \mu_{31} & \mu_{32} & \mu_{33} \end{bmatrix} \text{ and } \bar{\bar{\varepsilon}} = \begin{bmatrix} \varepsilon_{11} & \varepsilon_{12} & \varepsilon_{13} \\ \varepsilon_{21} & \varepsilon_{22} & \varepsilon_{23} \\ \varepsilon_{31} & \varepsilon_{32} & \varepsilon_{33} \end{bmatrix}, \tag{1.5}$$

Each element of these tensors is a complex number. In most composites, only the diagonal tensor elements are non-zero and the off-diagonal elements can be ignored. In this case a material may have six complex constitutive parameters. However, it is possible to have non-zero off-diagonal elements as well. For example, some ferrite crystals and gyrotropic plasmas are known to have off-diagonal permeability tensor elements. For the purposes of this book, we will consider only materials with zero-valued off-diagonals, as the development of methods for determining diagonal elements is still a rapidly evolving and active area of research.

Fortunately, when the permittivity and permeability tensors are diagonalized, the diagonal tensor components can be determined with independent measurements. For example, measurements of each component of ε and of μ are made by simply orienting the specimen so that the electric and magnetic fields of the incident wave correspond to the desired tensor components. This is possible in free-space measurement systems because the free-space propagation is a TEM mode and measurement fixtures are designed to have a linearly polarized incident beam. Of course, it is important to know the principal directions of the tensor within a material system. Otherwise, if the specimen is oriented at an angle that does not line up with the polarization direction, the resulting data will be a combination of two or more tensor components, making interpretation more complicated.

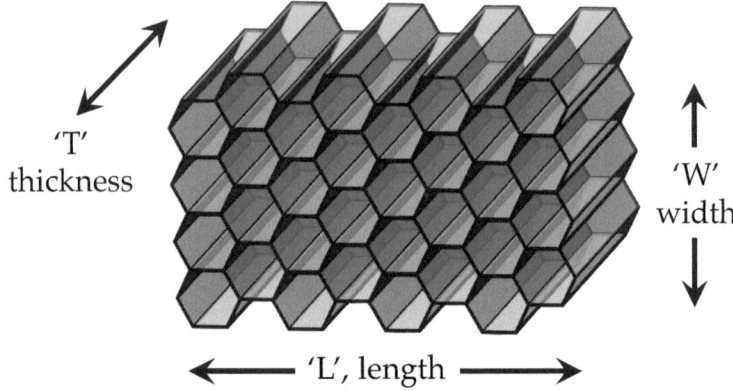

'T'
thickness

'W'
width

← 'L', length →

Figure 5   **Geometry of honeycomb core, an anisotropic material with three principal directions.**

HISTORICAL BACKGROUND

There is a long history of research in measuring electromagnetic materials leading up to the modern-day focused beam methods, and it is interesting to review this record. Some of the earliest known research in the interaction of free-space energy with materials was conducted by J.C. Bose in the 1890s [5]. During this time, Bose invented horn antennas, including waveguide-lens antennas, along with polarizers, prisms, and other components for manipulating microwave and millimeter wave energy. Bose's work was ground breaking, however interest in free-space characterization of materials did not occur in earnest until World War II, with the advent of the radar. Continued development of radar equipment and related RF and microwave technologies drove the need for understanding the properties of materials at these frequencies, whether they absorb, reflect, or transmit microwave energy. This importance stemmed from the need to incorporate materials in microwave components and antennas, as well as the desire to use materials for reducing the radar signatures of military vehicles.

An early treatise on free space measurement methods is found in the MIT Radiation Laboratory series published in the late 1940s [12]. This extensive reference documents the state-of-the-art in radar related technologies from that time period. In terms of material characterization, reference [12] describes methods for characterizing transmission and reflection of planar dielectric sheets illuminated by antennas in various configurations, and offers some techniques for inverting dielectric permittivity. Figure 6 shows drawings of early, pre-network analyzer fixtures for measuring the free space transmission phase (left) and amplitude (right) through a material specimen. Without the benefit of modern vector network analyzers, phase measurement required a micrometer for mechanically adjusting the distance between the two horn antennas. Additional details of these free space methods for obtaining permittivity are in [13], which includes research dating back to 1942. These references are restricted to dielectric properties of materials. However less than a decade later, these methods were generalized to magnetic property determination [14].

Figure 7 shows three different measurement configurations that were used during the 1940s for obtaining reflectivity of materials. These geometries enabled both normal incidence and oblique angle measurements of specimen reflectivity. The geometry on the right side of Figure 7 was pioneered at the Naval Research Laboratory and is also known as the "NRL arch" method [15]. It is estimated that NRL implemented this method in 1945 [16]. With a relatively simple set-up and modest cost, the NRL arch measurement method remains in common practice today.

The configuration in Figure 6 has also seen continued use in the form of an "admittance tunnel". Unlike the fixture in Figure 6, the more modern admittance tunnel contains the specimen and antenna(s) within an absorber-lined box. It derives its name from its original use in measuring the sheet impedance of thin conductive materials. In one version of the admittance tunnel, pictured in Figure 8 [17], the specimen is mounted at the end of the absorber-lined box and a moveable metal plate is placed behind the specimen. The metal plate position is varied to find the position producing the maximum and minimum in the reflection amplitude. Phase was also measured so that both the resistivity and the capacitance (or dielectric substrate permittivity) could be determined. A more prevalent admittance tunnel configuration placed the specimen between transmit and receive horn antennas. This type of tunnel is still in use today and is also called a transmission tunnel. It can be used to characterize thin sheet materials, as well as dielectric and magnetic slab specimens.

Two primary disadvantages exist with these early free-space tunnels and arches. The first is that practical limitations place the specimen near enough to the antennas to have significant phase taper across the specimen. This near-field illumination is a significant deviation from an ideal far-field plane wave, resulting in potential errors in the measured transmission and reflection coefficients. A more significant disadvantage however is the potentially large diameter of the illuminating beam. If a specimen is moved farther away from the transmit or receive antennas to compensate for near-field effects, the resulting beam diameter grows large enough that it may significantly illuminate the edges of the specimen. The subsequent edge diffraction then interferes with the specular reflection or direct transmission, resulting in significant amplitude and phase errors in the measured characteristics.

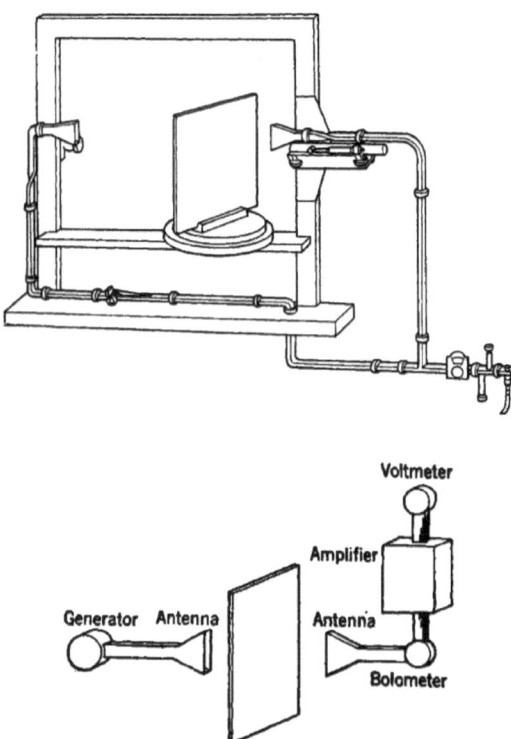

Figure 6   **Early configurations for free space transmission, including phase (top) and amplitude (bottom) from reference [1].**

While one way to minimize edge diffraction errors is to have a larger specimen size, this is often impractical. Another method, which is employed with some success in modern admittance tunnels, is the use of tapered apertures. While a tapered aperture can effectively impose a controlled amplitude taper on the illuminating microwave energy, it does not address the issue of phase curvature due to near-field effects.

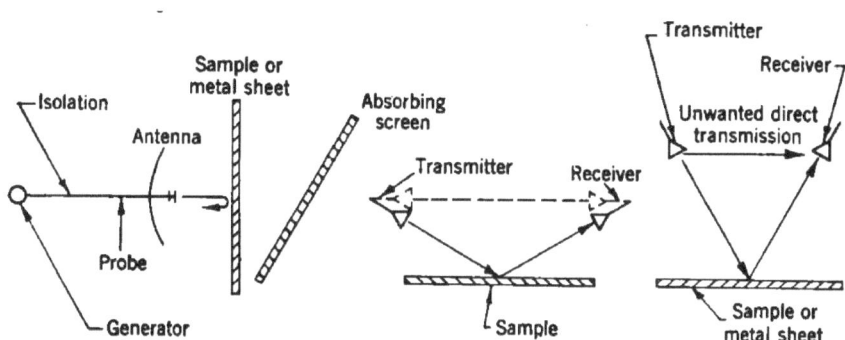

Figure 7   **Early configurations for free space reflection of materials from reference [1].**

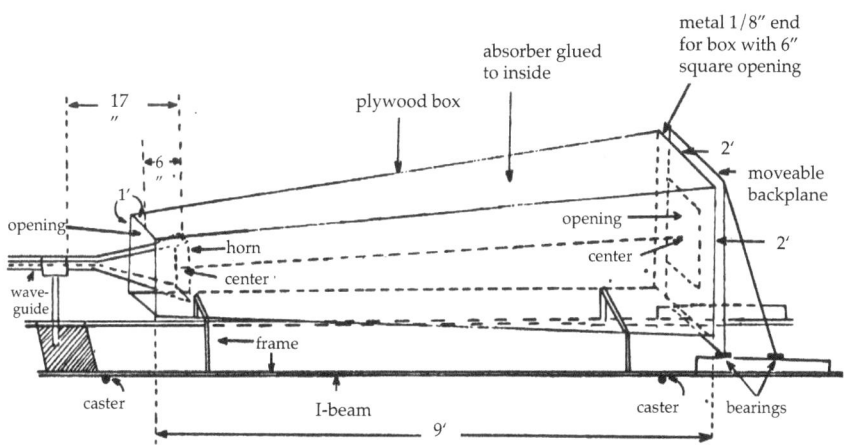

Figure 8   **Drawing of admittance tunnel that measures metal backed reflection [17].**

As early as 1950, researchers at NRL addressed this problem by incorporating dielectric lenses into horn antennas [18]. The need, in this case was to characterize the microwave properties of engine exhaust plumes, and the illumination pattern from standard horn antennas was too broad, resulting in a significant over-illumination of the plume. In this work, dielectric lenses were designed and built, successfully demonstrating the use of a focused beam to interrogate a finite-sized specimen. Similar, lens-focusing elements were used later to examine microwave properties of ionized trails behind hypersonic projectiles [19,20].

Also in the late 1940s and early 1950s, another parallel effort was underway to achieve the same result [21]. In this case, horn antennas combined with metal artificial dielectric lenses were used to correct the phase front for improved accuracy measurements of moisture content in bales of hay, and other agricultural specimens. Also in the same period, W. Culshaw developed a free-space measurement system that determined permittivity from bistatic reflectivity measurements at oblique incidence [22]. In this work, Culshaw used lenses to either collimate or focus the incident beam on smaller samples. His work demonstrated the improvement in accuracy that occurred by reducing edge diffraction errors.

With optical lenses as a design inspiration, it is not surprising that early focused-beam devices incorporated the direct dielectric analog. However subsequent researchers also developed alternate methods for focusing energy from a feed antenna. In 1961 Goubaou patented the idea of a beam-waveguide system that could use either dielectric lenses or parabolic reflectors to achieve beam focusing [23]. In 1966 Datlov, Musil, and Zacek applied focusing reflectors with horn antenna feeds to focus a beam used to characterize the microwave properties of a confined plasma [24]. A few years later, Bassett at Georgia Tech used four-foot diameter horn-fed ellipsoidal reflectors to focus microwave energy onto specimens that were heated to extremely high temperatures (2000 C) [25,26,27].

In the 1970s, Musil, Zacek et al. developed a unique alternative to a standard lens: a dielectric rod antenna [28]. This consisted of a dielectric rod inserted into the end of a horn antenna, which confined the radiated energy to a smaller area, much like a focusing lens. They successfully measured dielectric properties of materials with this device through direct contact of the dielectric rods with the specimen. This rod antenna concept has more recently been revisited with computational tools to improve the impedance match, and the rods are held a small distance away from the specimen rather than in direct contact [29]. The advantages of such a focusing mechanism include very small illumination area, similar to a very small focal length lens. The disadvantage of such an approach is that the proximity of the rod to the specimen reduces the convenience of high-temperature measurements or other applications where a longer standoff-distance between the fixture and the specimen is required. In both the dielectric rod antenna and in small focal length lenses, excessive focusing errors can occur with a beam diameter that is too small to satisfy the paraxial approximation. Focusing errors are discussed in greater depth in Chapter 5.

While there continues to be active development of systems with alternate focusing mechanisms, the dielectric lens remains the most broadly used method for microwave focused beam measurements. The lens' advantages include a relatively simple concept of operation, an ability to have a sizeable standoff distance between the lens and the specimen for extreme environments, and flexibility in accommodating a varied range of illumination areas (or focal lengths). Thus this book focuses on the implementation of lens-based focused beam systems. However many aspects of lens systems described here are also applicable to the other focusing mechanisms.

2. THEORY & IMPLEMENTATION

FOCUSED BEAM SYSTEM DESIGN

A focused beam system can be used to measure the scattered electromagnetic energy from a material specimen. From this measured scatter, the intrinsic properties of the specimen can be calculated, including parameters such as dielectric permittivity, magnetic permeability, sheet impedance, or diffuse and specular scattering performance. Similar to other transmission line methods, such as coaxial airline or waveguide methods, the focused beam system characterizes the scattering parameter matrix ("S-parameters") for a given thickness of the material specimen under test. The relationship between scattering parameters and the intrinsic properties of a material are derived in the next Chapter. For now, it is sufficient to know that the scattering parameters represent the signals that are transmitted or reflected by a material specimen when illuminated by an incident wave. Thus the first step towards determining the various material properties is to characterize the scatting parameters of a given specimen of that material. While the intrinsic properties of a material are usually independent of the size and shape of a material specimen, the scattering parameters are not, and they depend on geometrical factors such as thickness and incidence angle.

The concept of a focused beam system is primarily designed to provide a convenient mechanism for obtaining these scattering parameters with reasonable accuracy. A typical lens-based focused beam system geometry is shown schematically in Figure 9. Radiation from a feed antenna is directed to a dielectric lens, which focuses that radiation to minimized radius, where the slab specimen is placed. A reciprocal lens and antenna are on the other side to measure the transmission through the specimen. Alternately, the feed antenna and lens can collect the reflected energy from the specimen. The specimen consists of a flat sheet of material and by placing this specimen at the focus, a reasonable approximation to plane wave interaction is achieved.

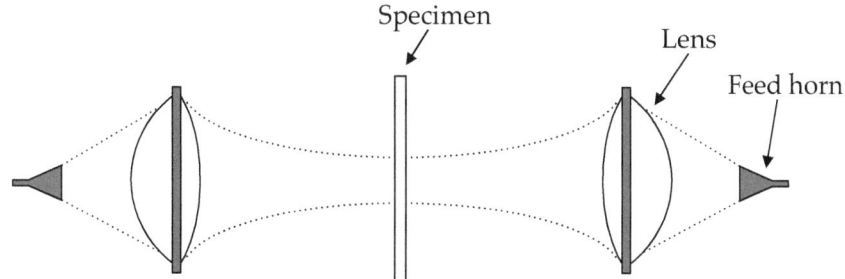

Figure 9   Geometry of a focused beam measurement system.

At optical frequencies, the implementation of lens systems is usually well described by geometrical optics, an approximation that is effective when the optical components are hundreds of wavelengths in dimension. However lens systems at microwave frequencies require the manipulation of wavelengths from millimetres to a large fraction of a meter. Practical considerations then drive the size of microwave optical components to be often less than ten wavelengths in dimension. Under these conditions, the tenants of geometrical optics become only approximate, and a more general theory is needed to describe microwave phenomena. In this quasi-optical regime, diffraction effects become important and Gaussian optical theory provides a convenient means for modeling the primary characteristics of a microwave focused beam system. More specifically, the geometrical optical theory of a lens' focus concentrates light into a single point, while Gaussian optics accounts for a minimum spot size due to diffraction. Gaussian beam theory as applied to microwave and millimeter wave systems is reviewed extensively elsewhere [30]. So this book gives only a brief outline of some of the pertinent aspects of the theory important for materials measurement. This combination of geometric optics and Gaussian optics are sufficient for design and characterization of lens systems as is shown in the following sections.

GAUSSIAN BEAM BASICS

In Gaussian optics theory, the radiation beam is assumed to have an amplitude taper described by a Gaussian function. For an axially symmetric, free-space beam propagating in the z-direction, the fundamental Gaussian mode is a solution to the paraxial wave equation, where $k_x, k_y \ll k$ and where time-harmonic fields with $\exp(i\omega t)$ are assumed,

$$\frac{\partial^2}{\partial x^2}u + \frac{\partial^2}{\partial y^2}u - 2ik\frac{\partial}{\partial z}u = 0 \qquad (2.1)$$

A solution to this equation is given by,

$$u = a(z)e^{\frac{-ik(x^2+y^2)}{2q(z)}} \qquad (2.2)$$

where $k = 2\pi/\lambda$ is the wavenumber, and a and q are two complex functions of z, that depend on the configuration of the beam. Such a beam is shown schematically in Figure 10, along with a few defined parameters. Plugging this solution into the paraxial wave equation, enables further evaluation of the unknown functions a and q. In particular q is called the "complex beam parameter" and can be written in terms of the parameters shown in Figure 10 [30],

$$q = z + iZ_R \qquad (2.3)$$

where the origin of z is defined to be at the minimum radius of the beam and,

$$Z_R = \frac{k}{2}w_0^2 \qquad (2.4)$$

is called the Rayleigh range (also sometimes called the confocal parameter). w_0 is called the beam waist and is the minimum radius, which occurs at the beam focus. The beam radius at other locations along the axis of propagation is given by,

$$w(z) = w_0\sqrt{1 + \left(\frac{z}{Z_R}\right)^2} \qquad (2.5)$$

   A convenient definition for specifying depth of focus is the Rayleigh range, Z_R. Rearrangement of Equation (2.5) shows that this Rayleigh range is the distance from the beam waist location to where the spot size has increased to $w(Z_R) = \sqrt{2}w_0$. In a material measurement apparatus, if a thick specimen extends outside of this range, then it will experience significant phase curvature, which may then reduce the accuracy of the plane wave assumption. Additionally, when a material specimen is placed in the focused beam, this Rayleigh range is reduced by the permittivity and permeability of the material since $k = k_0\sqrt{\varepsilon_r \mu_r}$ in Equation (2.4). Thus high index materials specimens should be thin enough to minimize focus depth errors within the constraints of the fixture design.

   Going back to our solution to the wave equation, Equation (2.2), and evaluating the function, $a(z)$ in terms of the above defined parameters while assuming that the on-axis amplitude at the beam waist is unity gives a more specific expression for the wave equation solution,

$$u(x, y, z) = \frac{w_0}{w(z)} e^{\left(\frac{-r^2}{w^2(z)}\right)} e^{\left(\frac{-ikr^2}{2R(z)}\right)} e^{i \tan^{-1}\left(\frac{z}{Z_R}\right)} \tag{2.6}$$

where $r^2 = x^2 + y^2$ is the distance from the axis. In this expression, we have also introduced another parameter,

$$R(z) = z\left(1 + \left(\frac{Z_R}{z}\right)^2\right) \tag{2.7}$$

which is the radius of curvature of the phase front as a function of position along the z-axis. At the beam focus ($z = 0$), $w_0 = w(z=0)$, and Equation (2.6) simplifies to

$$u(z = 0) = e^{\left(\frac{-r^2}{w_0^2}\right)} \tag{2.8}$$

   As Equation (2.7) indicates, the phase front at the focus has an infinite radius of curvature equivalent to a "flat" phase front. This flat phase front is key for enabling a good approximation of plane wave illumination at the specimen location. The inversion algorithms usually used in free-space methods assume plane wave interactions, so a flat phase front improves the accuracy of intrinsic property determination.

   In geometrical optics, the radius of curvature is an important parameter used to evaluate the transformation of optical rays from a point source through shaped reflectors or lenses. In the fundamental mode Gaussian beam equations above, there is an alternate expression for the complex beam parameter that is analogous to this geometric radius of curvature,

$$\frac{1}{q} = \frac{1}{R} - i\,\frac{2}{kw^2} \qquad (2.9)$$

From this expression, we see that q is effectively a complex radius of curvature, and this provides a convenient mechanism for transforming Gaussian beams through optical elements such as lenses using a similar formalism to that applied in geometrical optical design.

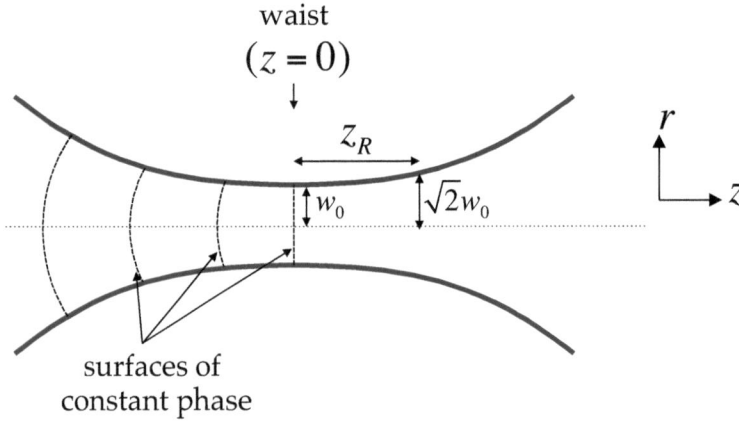

Figure 10   **Features of a focused Gaussian beam.**

The Gaussian assumption is also useful because it provides straightforward expressions for describing the performance of microwave lens systems. For example, integrating the power density, $|u|^2$, in cylindrical coordinates gives the fraction of power outside a radius, r_1 at an arbitrary z,

$$\Delta P(r > r_1) = e^{-2r_1^2/w^2} \qquad (2.10)$$

This quantity is also the same as the relative power density at that same radius. Equation (2.10) is useful for the design of a measurement fixture since it provides guidance on the minimum specimen size. For example, a common rule of thumb for minimum lateral specimen dimensions (in the x and y directions) is that they should be no smaller than the -20 dB radius of the beam, which is just over one and a half times the beam waist. In this case, the fractional power going around the specimen will be no more than 1%. However if a specimen has a very high insertion loss, then a larger size is required to reduce the fractional power outside the specimen from dominating the transmitted signal. Similarly, Equation (2.10) provides an estimate for the required radius of the lens, since the lens must i) be large enough to encompass the energy radiated from the feed antenna and ii) be large enough to achieve the desired focused spot for the specimen illumination.

As a theoretical framework for describing the interaction of microwave energy with lenses and planar sheets, Gaussian beam theory provides a reasonable approximation for the design of a focus beam measurement system. However, the paraxial approximation assumed by this theory is still an approximation, and can ultimately limit the accuracy of the plane wave assumption. Furthermore, when measurement hardware is built, deviations from this approximation may require experimental adjustment of the system, often by repositioning the relative positions of the feed antennas, lenses, and specimen, to account for deviations from ideal assumptions.

With that in mind however, the Gaussian beam and geometric optics approximations are still sufficient to determine an optimum lens shape for a given set of design constraints, and measurement examples shown later in this book are based on a system designed with these approximations. The following section provides a detailed description of the theory that can be used to design lenses in a focused beam material measurement system.

LENS DESIGN

The first step in designing a focusing lens is to have a model of the feed antenna in terms of the Gaussian parameters described above. In particular we translate typical antenna parameters, such as the 3-dB beam width, into an equivalent Gaussian beam representation, and start with expressing Equation (2.10) in terms of angle. Assuming the paraxial approximation, we can estimate that $r_1^2/w^2 \approx \theta_{3dB}^2/\theta_{1/e}^2$, where $\theta_{1/e}$ is the angle from the axis to where the field amplitude is down to $1/e$, and θ_{3dB} is the half-angle that defines the 3 dB beam width. If we know the 3 dB beam width of the feed antenna, we calculate the $1/e$ angle by knowing the power at that angle is ½ of the power at the beam center, $e^{-2\theta_{3dB}^2/\theta_{1/e}^2} = 1/2$. This results in $\theta_{3dB} = 0.589\,\theta_{1/e}$.

Furthermore, with the assumption that a feed antenna has a Gaussian beam emanating from it, we can calculate the equivalent waist of that beam from the $1/e$ far-field beam width. The growth of the $1/e$ radius of the beam can be defined in terms of an angle from the origin (i.e. the beam waist location or the phase center of the antenna), $\theta = \tan^{-1}(w/z)$. In the far field ($z \gg Z_R$), Equation (2.5) can be rearranged to show that $w/z \to w_0/Z_R$. Then using the paraxial approximation along with Equation (2.4) we can express the far-field divergence angle in terms of the beam waist,

$$\theta_{1/e} = \tan^{-1}\left(\frac{\lambda}{\pi w_0}\right) \approx \frac{\lambda}{\pi w_0} \tag{2.11}$$

Once we have these characteristic parameters of the feed, we can then determine a design for the lens that is optimized for that feed. The design of a typical focused beam measurement system uses a bi-convex lens to transform the source radiation from the feed antenna into a focused beam, and geometrical optics design approximations can be employed to determine the lens shape. In microwave lenses [30,31,32], the usual design concept is that each side of the bi-convex lens transforms rays between the divergent radiation from a focus and an approximate plane wave at the center of the lens. Thus one side of a lens is designed based on the source antenna, and the other side of a lens is designed based on the desired focused beam size at the specimen.

This concept of converting a diverging beam to a collimated beam is shown in Figure 11, and is applied in the lens design equations that follow. For simplicity, the following discussion applies specifically to the "input" half-lens which transforms the antenna radiation into an approximately collimated beam at the aperture plane of Figure 11. An analogous set of equations can then be used design the "output" half of the lens.

The shape of the lens is derived by applying geometrical optics, or more specifically Fermat's principal. This principal states that the total transit time for a ray emanating from a point source and travelling through the lens must be the same no matter what path that ray takes through the lens. Assuming rotational symmetry and using the geometry shown in Figure 12, a formula can be derived to describe the thickness as a function of radius. For this calculation, the radiating antenna is assumed to be a point source that is located at the origin, O, and at a distance from the apex of the lens that is defined as the focal length, f_i. Assuming rotational symmetry, X is then the distance from the origin to any point on the surface of the lens. The lens has an index of refraction, n, and is assumed to have negligible loss.

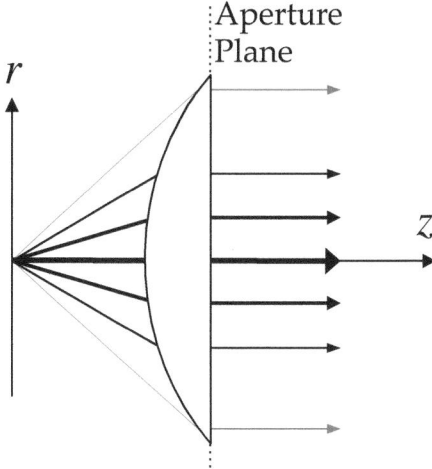

Figure 11 **Sketch of half-lens showing the principal of transforming
from a point source to a collimated beam.**

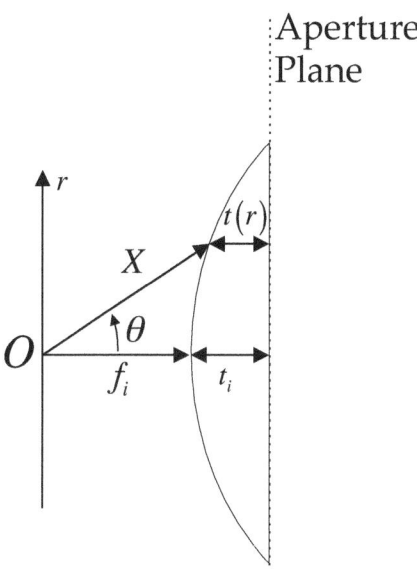

Figure 12 **The parameters that define the lens shape based on
geometric optics.**

In order to collimate the diverging rays from the origin, the electrical distance from O to any point along the aperture plane must be a constant. When the angle, θ is zero, then the electrical length between the origin and the aperture plane is given by,

$$Electrical\ length(\theta = 0) = f_i + nt_i \tag{2.12}$$

where t_i is the maximum thickness of the lens. Straightforward geometry then enables us to calculate the electrical length at other angles,

$$Electrical\ length(\theta) = X + nt = \sqrt{r^2 + \left(f_i + (t_i - t)\right)^2} + nt \tag{2.13}$$

where t is the thickness and is a function of r. By comparing these two equations, we can then determine the shape of the lens,

$$t(r) = t_i + \frac{f_i - \sqrt{f_i^2 + \left(\frac{n+1}{n-1}\right)r^2}}{n+1} \tag{2.14}$$

The shape of the output lens that focuses at a given focal length is also given by a similar equation.

An alternate expression for the shape of the lens in terms of X is given by similarly evaluating the geometry shown in Figure 12 along with Fermat's principal stated another way: $X + nt = f_i + nt_i$. Noting that $\sin\theta = r/X$ and that $\tan\theta = r/(f_i + t_i - t)$, then

$$X = \frac{(n-1)f_i}{n\cos\theta - 1} \tag{2.15}$$

For $n > 1$, this defines a hyperbola of revolution, where the focal point of the lens is at the focus of the hyperbola.

In Cartesian coordinate systems, the standard expression for a hyperbola is given by,

$$\frac{x^2}{a^2} - \frac{y^2}{b^2} = 1 \tag{2.16}$$

In the coordinates defined in Figure 11, $x \to z$ and $y \to r$. Recognizing from Figure 12 that $X^2 = z^2 + r^2$, and after some algebra, the standard hyperbolic form for the lens defined by Equation (2.15) is given by the following equation,

$$\frac{z^2}{a^2} - \frac{r^2}{b^2} = \frac{z^2}{\dfrac{f_i^2}{(n+1)^2}} - \frac{r^2}{\dfrac{f_i^2(n-1)}{(n+1)}} = 1 \qquad (2.17)$$

Now that we have an expression for the lens shape based on geometrical optics, we then propagate a Gaussian beam through that lens and iterate as needed for effects due to the Gaussian beam phenomena. The position and diameter of the lens must account for the radiated pattern of the feed antenna as well as the desired beam waist at the focal point where the specimen is placed. More practically the lens diameter is usually limited by cost and space constraints and this ultimately sets the lower-frequency limit of the designed lens.

Assuming a fixed lens diameter, the antenna position is then set so that the antenna illuminates the full lens area efficiently but without over-illuminating. Otherwise the energy on the other side of the lens will contain unwanted diffraction effects. A first estimate of the antenna-lens spacing can be determined by setting a criterion for the level of power at the lens edge (e.g. the power at the edge is no more than -20 dB from the center). By calculating the distance, z, from the feed antenna to the lens position so that the power at the lens' outer radius equals the design criteria, we have an initial estimate of the separation between the antenna and the lens.

However, this initial estimate will probably be inaccurate since the thickness of the lens for this application is likely to be a significant fraction of the total lens-antenna separation. This 'thick lens' effect effectively causes the beam width exiting the lens to narrow due to refraction by the lens. Thus we need to derive an expression for propagating the Gaussian beam though the lens that accounts for this thickness effect.

Gaussian beam propagation through a lens can be characterized in terms of an ABCD matrix formalism (also known as a ray transfer matrix) that is widely used in geometrical and Gaussian beam analysis [33, 34]. The ABCD matrix allows the output of an optical element to be written in terms of its input. An input ray is characterized by a vector with components r and θ as defined in Figure 12. The ABCD matrix is than applied to determine the vector that represents the output ray,

$$\begin{pmatrix} r' \\ \theta' \end{pmatrix} = \begin{pmatrix} A & B \\ C & D \end{pmatrix} \begin{pmatrix} r \\ \theta \end{pmatrix} \quad \text{or} \quad \begin{matrix} r' = Ar + B\theta \\ \theta' = Cr + D\theta \end{matrix} \qquad (2.18)$$

where the primed components are for the output beam. The coefficients A, B, C, D, are determined by the specific properties of a given optical element (e.g. a lens, a mirror, or an interface between different media).

Because the paraxial approximation is assumed, $\tan\theta \approx \theta$, which is the slope of the ray under consideration. The radius of curvature of the ray is then $R \approx r/\theta$, and the above transformation can then be restated in terms of R,

$$R' = \frac{AR + B}{CR + D} \tag{2.19}$$

Generalizing this equation to the case of a Gaussian beam approximation, the geometrical radius of curvature is simply replaced by the corresponding Gaussian radius of curvature (i.e. the complex beam parameter) as defined in Equation (2.9),

$$q' = \frac{Aq + B}{Cq + D} \tag{2.20}$$

To use this formulism to evaluate a specific lens design, we must have a library of different ABCD matrices to apply. For the simplest case of a length, l, of free space, it is straightforward to show that this ABCD matrix is,

$$M_1 = \begin{pmatrix} 1 & l \\ 0 & 1 \end{pmatrix} \tag{2.21}$$

This same matrix also applies in a homogeneous medium of arbitrary index.

For a planar interface between to regions with difference indices of refraction, the ABCD matrix is,

$$M_2 = \begin{pmatrix} 1 & 0 \\ 0 & n_1/n_2 \end{pmatrix} \tag{2.22}$$

where the refractive index of the medium the ray is coming from is n_1 and the index of the medium the ray is going towards is n_2. This second matrix can be verified by applying it to Equation (2.18) and showing that it satisfies Snell's law for refraction (when the paraxial approximation is applied).

An often-used transformation is for the thin lens, which is a focusing element consisting of one or two curved interfaces and where the physical separation and thickness is neglected. The equation that describes the thin lens transformation is given by [34],

$$M_{thin\ lens} = \begin{pmatrix} 1 & 0 \\ \dfrac{-1}{f} & 1 \end{pmatrix} \tag{2.23}$$

where f is the focal length of the lens. Unfortunately, the typical design for a focused beam material measurement system requires a lens with a thickness that is a significant fraction of the separation between the antenna and the lens output. This thin lens formula ignores the effect of that thickness, so is too approximate to be particularly useful in microwave measurement systems.

In the lens design described here, we have a hyperbolic shaped surface that can be described by the following ABCD matrix, [35, 36]

$$M_3 = \begin{pmatrix} 1 & 0 \\ \dfrac{-1}{an(n+1)} & n \end{pmatrix} \tag{2.24}$$

where a is the semi-major axis of a hyperbola. For the hyperbolic shape we developed above, we have already obtained a relationship between this parameter and the input parameters of the lens shape. Specifically, from Equation (2.17) we can write, $a = f_i/(n+1)$.

We are evaluating a half lens, which includes a hyperbolic surface, a distance within the lens medium, and a flat transition from the lens back to air. So to construct a matrix that includes these three parts we simply cascade together the appropriate matrices for each of these parts of the system. Because the output vector of the beam parameter is on the right of this cascade, the matrices are placed in order from right to left rather than left to right,

$$M_2M_1M_3M_1 = \begin{pmatrix} 1 & 0 \\ 0 & n \end{pmatrix}\begin{pmatrix} 1 & t_i \\ 0 & 1 \end{pmatrix}\begin{pmatrix} 1 & 0 \\ \dfrac{-1}{an(n+1)} & n \end{pmatrix}\begin{pmatrix} 1 & f_i \\ 0 & 1 \end{pmatrix}$$
$$= \begin{pmatrix} 1 - \dfrac{t_i}{f_i n} & f_i \\ \dfrac{-1}{f_i} & 0 \end{pmatrix} \tag{2.25}$$

where the ABCD matrices multiplied above correspond to the distance from the feed antenna to the lens (M_1), the hyperbolic lens face (M_3), the thickness of the lens (M_1), and the flat face of the lens (M_2). These regions are shown in Figure 13, along with the antenna to lens face distance, f_i, the lens thickness t_i, and the refractive index, n. This matrix is then applied to the beam emanating from the antenna to obtain the beam parameter at the output plane of the lens,

$$q_{output} = \frac{\left(1 - \frac{t_i}{f_i n}\right) q_{antenna} + f_i}{\left(\frac{-1}{f_i}\right) q_{antenna}} \tag{2.26}$$

Another way to derive the effect of the lens on the beam emanating from the antenna is to directly map the incident field amplitude from one side of the lens to the other side while accounting for the non-zero z-depth (Figure 14) of the convex side. Stated another way, the incident field amplitude on one side of the lens, multiplied by the differential area on that side must be mapped to the equivalent field amplitude and differential area on the planar side of the lens,

$$u(\theta)\delta\theta = u'(r)\delta r \tag{2.27}$$

where u represents the field amplitude on the convex side and u' is the field amplitude on the planar side of the lens. This is shown schematically in Figure 14.

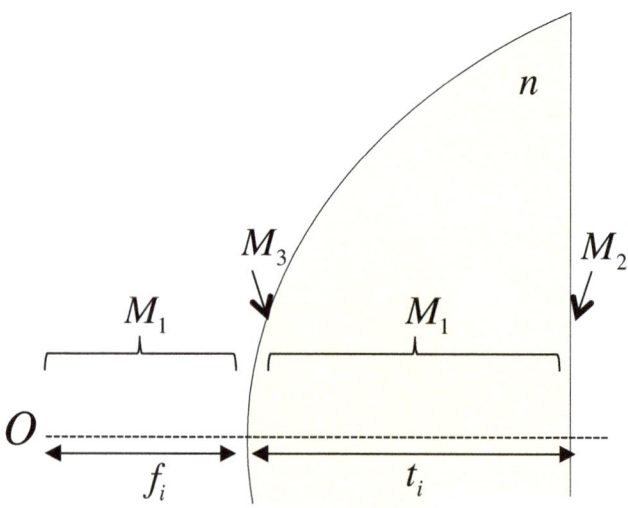

Figure 13   **Geometry of lens showing the regions corresponding to different cascaded ABCD matrices ($M_{total} = M_2 M_1 M_3 M_1$) of Equation (2.25).**

Assuming rotational symmetry, we can rewrite $\delta\theta$ and δr as infinitesimal line segments in spherical coordinates ($\delta\theta = Xd\theta$ and $\delta r = dr$) and reformulate this equation to obtain the unknown field amplitude profile on the planar side of the lens in terms of the field on the convex side,

$$u'(r) = X\frac{u(\theta)}{dr/d\theta} \tag{2.28}$$

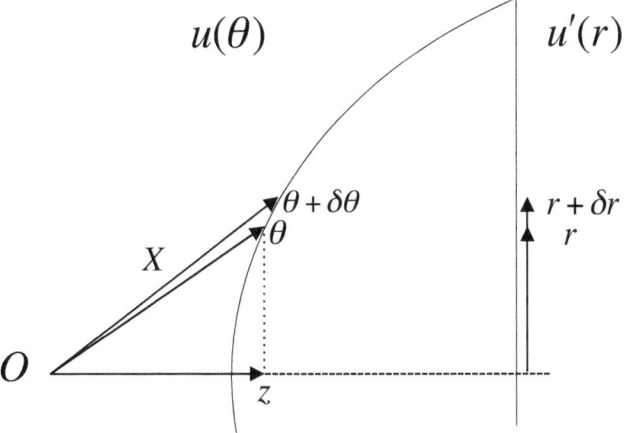

Figure 14 **Schematic showing relevant variables for the translation of a beam profile from one side to the other of a half-lens.**

From Figure 12, we know that $r = X\sin\theta$ and from Equation (2.15) we know the angular dependence of X, thus we can evaluate the derivative,

$$\frac{dr}{d\theta} = \frac{d}{d\theta}\left(\frac{\sin\theta\,(n-1)f_i}{n\cos\theta - 1}\right) = \frac{X(n-\cos\theta)}{n\cos\theta - 1} \tag{2.29}$$

Because $u(\theta)$ is the field amplitude on the hyperbolic surface it varies with z. Using the geometry of Figure 14, we can rewrite the expression that defines the hyperbolic surface so that it is a function of z,

$$X = \frac{(n-1)f_i}{n\frac{nz}{X} - 1} = f_i(1-n) + nz \tag{2.30}$$

After some algebra, this can be rearranged to obtain an expression for z that follows the hyperbolic surface as a function of r,

$$z = \frac{nf_i}{n+1} + \sqrt{\frac{r^2}{n^2 - 1} + \frac{f_i^2}{(n+1)^2}} \tag{2.31}$$

Using Equation (2.30), we can also rewrite the expression for the transformed field amplitude in terms of this z that defines the hyperbolic surface,

$$u'(r) = \frac{-f_i}{nf_i - (n+1)z}u(\theta) \tag{2.32}$$

The expression for z can then be inserted into Equation (2.6) to determine $u(\theta)$ on the surface of the hyperbolic lens, and this in turn is inserted into Equation (2.32) to calculate the new field distribution at the output aperture of the half-lens.

To illustrate the use of these equations for a lens design, Figure 15 shows a plano-convex lens, optimized to provide a collimated beam from a feed antenna with a 50-degree 3-dB beam width. In this design, the permittivity of the lens was assumed to be 2 and the maximum radius of the lens was 70 cm. The design of this lens began with the feed antenna, which was characterized in terms of an equivalent Gaussian beam waist. The distance of the antenna relative to the lens was optimized so that the power level of energy exiting the lens at the maximum radius was approximately -20 dB from the center of the beam. Once this focal distance was established, then Fermat's principal was used to calculate the convex shape. Figure 15 also shows the $1/e$ beam radius as a function of z, since this corresponds to the Gaussian beam radius defined in Equation (2.5). Note that both the $1/e$ and 3-dB beam radii are reduced after the energy is refracted through the plano-convex lens.

The calculated field profile at the output side of the lens is shown in Figure 16. The profiles shown in this plot were calculated by three different methods: ABCD matrix method, Fermat's principal, and numerical calculation with a full-wave computational code. The ABCD matrix method and Fermat's principal curves were calculated with equations described earlier in this chapter. The Finite Difference Time

Domain (FDTD) curve was calculated by modeling a two-dimensional representation of the lens. A distribution of plane waves was injected with appropriate weights to provide the Gaussian beam emanating from the feed antenna. The fields adjacent to the output plane of the lens were then sampled and plotted. Because the ABCD matrix and Fermat's calculations are approximate, they disagree somewhat with each other and with the FDTD calculation. There is also a slight amount of ripple in the FDTD simulated profile due to multiple reflection effects from the lens/air interfaces.

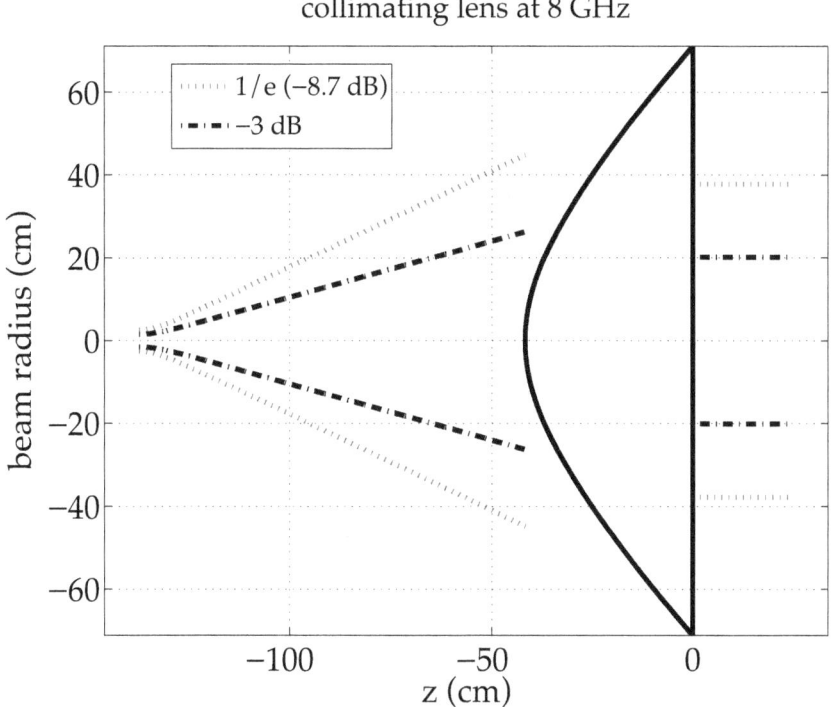

Figure 15   Example collimating lens design showing shape and calculated beam widths when optimized at 8 GHz for a 50-degree beam width feed antenna.

Full-wave electromagnetic simulations such as the FDTD simulations shown here are time-consuming, and may be practical only for verifying the design performance. Fortunately, the approximate methods described above provide sufficient accuracy to achieve most design goals.

In addition, the actual feed antennas typically deviate somewhat from the idealized Gaussian beam representation across frequency. In other words, the phase center of many wide-band feed antennas can change with frequency, so that antenna-lens distance adjustments may be necessary once a focused beam system is built.

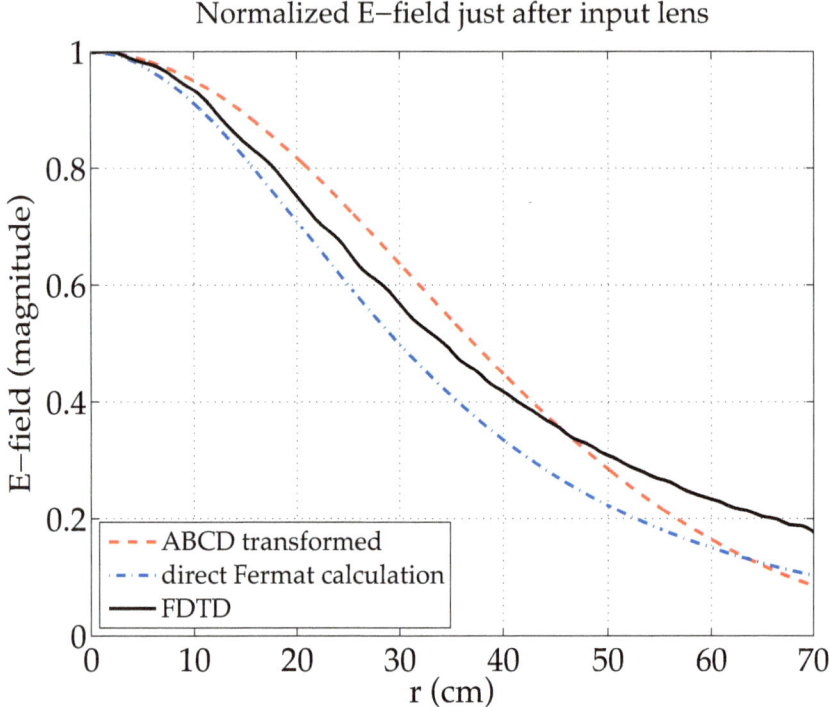

Figure 16   **Output beam profile for an example-collimating lens calculated by three methods: ABCD matrix method, Fermat's principal, and Finite Difference Time Domain (FDTD) full-wave simulation code.**

Lens System Construction

Lenses are typically manufactured from dielectric materials such as a polystyrene, Teflon®, or tooling foam, which have low-to-moderate dielectric permittivities. The dielectric constant must be high enough to provide sufficient phase delay for manipulating the electromagnetic energy, but low enough to keep reflections at the lens/air interface to a manageable level.

Feed antennas for focused beam measurement systems are usually horn antennas with linear polarization. Dual polarization antennas can also be used and are convenient for characterizing anisotropic specimens, but may increase uncertainties when measuring strongly anisotropic materials. In principal, the best accuracy is obtained with antennas that have good polarization purity and a Gaussian-like beam pattern. Selection of an appropriate feed also depends on the wavelength band of interest. With a fixed physical aperture, the gain characteristics of a typical feed antenna are proportional to A/λ^2, where A is the area of the antenna aperture. When transformed into a focused beam with a bi-convex lens, this results in a beam waist that is proportional to wavelength (or inversely proportional to frequency). Thus the minimum size of the specimen under test is driven by the lowest frequency of interest. Because of this frequency dependence, a customary way to specify the beam of a focused beam system is in terms of a frequency scaled parameter, k_0w_0, where $k_0 = 2\pi/\lambda_0$ is the wave number in free space.

A commonly used beam configuration used in a number of measurement laboratories is $k_0w_0 \approx 8$. In this case, the beam waist varies from 19 cm at 2 GHz to 2 cm at 18 GHz. An example of this configuration is shown in the photograph in Figure 17. Following the -20 dB rule-of-thumb for minimum lateral specimen size and using Equation (2.5), a minimum specimen for this system should be no smaller than 58 cm across at 2 GHz, or 28 cm across at 4 GHz. In the bi-convex lenses shown in Figure 17, each lens is actually constructed from two half-lenses held together by a metal ring and mounted on sliding rails. This flexible design enables different lens faces to be switched out for different waist sizes. For example, measurement of inhomogeneous materials at high frequencies can require a larger illumination area to encompass the characteristic length scales in the material. Replacing one of the lens halves with a different focal length face then allows reconfiguration of the waist size. The lenses in Figure 17 are machined from Rexolite®, which is a cross-linked polystyrene plastic. Rexolite has a highly stable dielectric constant from UHF through millimeter wave frequencies.

Another example of a focused beam system is pictured in Figure 18. In this system, the lenses were manufactured from polypropylene, with a dielectric permittivity of 2.28. To build this apparatus, large blocks of polypropylene were cast and then machined on a vertical bore mill. The large, 6-foot (183 cm) diameter of these lenses allows measurement at frequencies as low as 500 MHz.

Figure 17   Photograph of focused beam measurement system with 2-foot (61 cm) diameter lenses.

Figure 18   Photograph of UHF focused beam system with 6-foot (183 cm) diameter lenses.

CALIBRATION

Determining intrinsic dielectric or magnetic properties of a material specimen with a focused beam involves several steps: calibration, time-domain filtering, and property inversion. Calibration establishes the quantitative scattering parameters of a specimen by comparing measured data to appropriate reference standards. Time-domain filtering minimizes systematic errors caused by multi-path reflections within the fixture hardware. Property inversion determines the desired dielectric and magnetic properties of the specimen, usually through numerical solution of the equations that relate scattering parameters to the intrinsic properties. These various procedures are outlined in this section.

The first step to characterizing materials in a focused beam system involves quantitative measurement of the network scattering parameters. The focused beam system works by simulating a plane-wave that interacts with the specimen, which is a planar slab of material. For homogeneous materials, this planar slab can be considered a two-port network, defined by a two by two matrix of scattering parameters to be measured (transmission and reflection in forward and backward directions). For this reason, modern free-space measurement systems utilize automated network analyzers to acquire scattering parameter data.

The raw data from the microwave receiver must be calibrated to obtain quantitative scattering parameters. These data are complex having both an amplitude and phase (or alternately real and imaginary components) associated with each data point. The measured phase corresponds to the time delays of the focused beam system with and without specimens. This is different than the phase curvature in the Gaussian beam discussion earlier in this chapter. For reflection data (S_{11} or S_{22}), a very simple calibration is achieved by dividing the data from the specimen under test with a separate measurement of a short (reflection from a metal plate). For transmission data (S_{21} or S_{12}), a simple calibration is made by dividing the specimen data with a measurement of a thru (transmission with no specimen). Because of the thickness of the specimen, phase corrections must also be applied to the measured signals. These phase corrections correspond to the signal path in the focused beam system displaced by the specimen thickness.

For example, with S_{11} measurements, the front face of the shorting plate face is usually in the same position as that of the material specimen. So no phase correction is needed. Hence the calibrated S_{11} is just the ratio of the specimen reflection (S_{11}) to the metal plate or short reflection (R_{11}),

$$S_{11}^{cal} = \frac{S_{11}^{sample}}{R_{11}^{short}} \qquad (2.33)$$

If the specimen is placed in front of the shorting plate position (i.e. so that the back of the specimen is in the same position as the front of the short plate), then Equation (2.33) is multiplied by a phase correction,

e^{-2ik_0t}. This correction corresponds to the path length that is displaced by the specimen.

For S_{21} measurements, the signal path without a specimen has an extra length corresponding to the thickness of the specimen, t. Hence a phase correction is included that corresponds to this extra length,

$$S_{21}^{cal} = e^{-ik_0t} \frac{S_{21}^{sample}}{R_{21}^{thru}} \quad (2.34)$$

The simple calibration equations (2.33) and (2.34), use only a single calibration standard for each S-parameter, and are termed 'response only' calibrations. A more rigorous calibration method uses two different calibration standards and is called a 'response and isolation' calibration. The calibrated scattering parameters for this method are obtained by,

$$S_{ij}^{cal} = \frac{S_{ij}^{measured} - I_{ij}}{R_{ij} - I_{ij}} \quad (2.35)$$

where ij is the set of scattering parameters (11, 12, 21, 22). R_{ij} are the same reponse standards already discussed (metal plate for reflection and no specimen for transmission). In addition isolation standards, I_{ij}, are used, consisting of an opaque metal plate for transmission and a matched load for reflection.

In the case of reflection, this response and isolation calibration method is similar to radar cross-section (RCS) measurements. In RCS measurement ranges, a radar is used to illuminate a target, and the backscatter from that target is measured. Whether RCS measurements are done in an outdoor facility or in an indoor anechoic chamber, there are always undesired clutter sources in the vicinity of the target under test, which contribute to the total measured signal. An RCS measurement calibration utilizes a background measurement that occurs in the absence of the target under test, and which is subtracted from the target data to minimize the unwanted clutter signals. In the focused beam system, the isolation calibration is akin to this RCS background subtraction, where the focused beam "background" consists of reflections from discontinuities in the transmission path, such as the lens faces, the feed antenna, and the network analyzer ports.

This RCS measurement analogy applies to the response calibration standard as well. In an RCS range, a measurement of a known standard target (such as a sphere or cylinder) is also conducted, which provides a mechanism for quantifying the unknown target scatter by accounting for the power levels of the transmitter, the sensitivity of the receiver, and geometrical considerations in a given RCS measurement facility. This known target measurement is akin to the response standard in the focused beam calibration method, which is used to normalize the unknown specimen response to account for transmitter power, receiver sensitivity, and transmission line losses in the focused beam system

When accurate reflection measurements are needed, the position of the specimen should exactly correspond to the position of the reflection response standard (metal plate). Any deviation between the positions of the two will result in a phase error. These errors can also arise from warped specimen geometries, where slight curvature occurs due to material flexibility or internal material stresses. Specimen curvature results in a net position offset of the middle of the specimen from the reflection calibration plane, causing a phase error in the measured reflection coefficient. However, additional procedures can be used to account for these phase errors.

For homogeneous materials, the reflection and transmission should be identical in both directions. Normally only one transmission scattering parameter (e.g. S_{21}) and one reflection scattering parameter (e.g. S_{11}) are measured to determine the complex permittivity and permeability. Measuring the scattering parameters in the other directions provides redundant data. However when the specimen location deviates from the reference plane position, equal but opposite phase offsets occur in the forward and backward reflection directions (S_{11} and S_{22}). Therefore, measuring all scattering parameters can provide information on this phase error, which can then be corrected in the inversion algorithm.

Thus this 'four-parameter' method has the advantage of not needing a precise placement of the specimen relative to the calibration reference plane. In particular, the phase error associated with specimen placement is eliminated, resulting in greater accuracy for the permittivity and permeability calculations. The following procedure can be used to obtain the response and isolation calibration coefficients in this method:

1) Measure S_{21} and S_{12} (i.e. thru or fixture with no specimen) to obtain transmission response coefficients (R_{21} and R_{12}).
2) Insert metal plate with known thickness t_m and measure S_{11} and S_{22} (R_{11} and R_{22}) *without moving the plate* to obtain reflection response coefficients.
3) Leave plate in place, measure S_{21} and S_{12} to obtain transmission isolation coefficients (I_{21} and I_{12})
4) Remove metal plate and insert broad-band foam absorber, offset from specimen position as follows:

a. To measure reflection isolation in the 11-direction (for I_{11}), place the back of absorber against far lens so that it is outside of the time-domain gate – at least 30 cm beyond specimen position and measure reflection isolation.

b. For I_{22}, place the back of absorber against the opposite lens so that it is also beyond the time domain gate/specimen position and measure reflection isolation in the 22-direction.

Once these calibration measurements are complete, then the specimen is inserted and all four scattering parameters are measured without moving the specimen. The calibrated scattering parameters for the specimen are then obtained with equation (2.35). The calibration must also account for the transmission line displaced with the metal calibration plate by including the calibration plate thickness, t_m as well as the specimen thickness, t_s.

For most applications, the four-parameter response and isolation calibration method provides sufficient accuracy. However, there may be occasions when a more extensive calibration method is desired. One such method that has been implemented [37] is the TRL or Thru-Reflect-Line calibration. The TRL calibration also uses the metal plate and thru standards of the simple response method. However instead of the isolation standards of the Response and Isolation method, TRL uses 'line' standards. The line standards consist of extra sections of transmission line, which impose a phase shift on the measured transmission. Ideally the phase shift should be approximately 90 degrees. When wide-band feed antennas are used (e.g. 2-18 GHz), this dictates that multiple line standards are needed to avoid the ambiguity of a 180-degree phase shift somewhere within the measurement band. In a free-space measurement, the line standard is implemented by physically increasing the separation between the two sides of the focused beam system.

While the TRL calibration method can give better results than a Response and Isolation method, it requires moving the feeds when used in a free-space system, which requires additional hardware (e.g. an accurate linear translation mechanism). Furthermore, even with a TRL calibration, there remain significant residual errors due to multi-path reflections, and time domain filtering is usually applied to further reduce measurement ripple. When a time-domain filter is used, the accuracy difference between

TRL and Response and Isolation calibration methods becomes much less significant.

TIME DOMAIN PROCESSING

When data is acquired over a reasonably wide range of frequencies, a Fourier transform can be applied to convert frequency-domain data into time-domain data. This enables the separation of the measured signals into different components from the unknown specimen and from other discontinuities within the measurement system. Data measured with a network analyzer is discrete in frequency so a discrete transform is required to view the data in the time domain. The resolution in the time domain is proportional to the bandwidth of the frequency domain, $\Delta t = 1/(f_{max} - f_{min})$. The unambiguous range of the time domain is proportional to the number of points, N, in the frequency domain, $t_{unambiguous} = (N - 1)\Delta t$. Therefore, using a high number of points is preferred so that the unambiguous range is large enough to avoid aliasing (overlap) with other undesired signals.

Examples of time-domain data are in Figure 19, which shows the reflection amplitude measured with a focused beam system after a Fourier transform. In both plots, the Fourier transform was performed over an 8 to 12 GHz band. The blue dotted line of each plot shows the measured response when a tilted absorber is placed just beyond the specimen position and the black solid curve shows the response when a normal incidence metal plate is placed at the specimen position. Peaks are labeled according to their source. The mismatch due to the feed antenna is evident at 8 ns. For the quad-ridge horn, the feed mismatch is only a few dB lower than the reflection of the metal plate placed at the specimen position (just before 20 ns). For the standard gain horn, the match is significantly lower -- with the horn reflection 20 dB below the metal plate. Additional multipath reflections are evident at multiples of the antenna-specimen separation.

Calibration will effectively reduce the mismatch peaks from the cable and feed antenna mismatches. However, the multipath reflections depend on the amplitude attenuation and phase delay introduced with the specimen, so they can still be significant, even after calibration. Thus a filter is usually applied to selectively minimize all the time-domain peaks except for the specimen signal. In time-domain, this filter is a window that is multiplied with the data to preserve the desired signal while minimizing other signals at other times. A number of different window shapes exist for filtering discrete signals, and a window that works particularly well for the signals of interest here is the Kaiser-Bessel window [38].

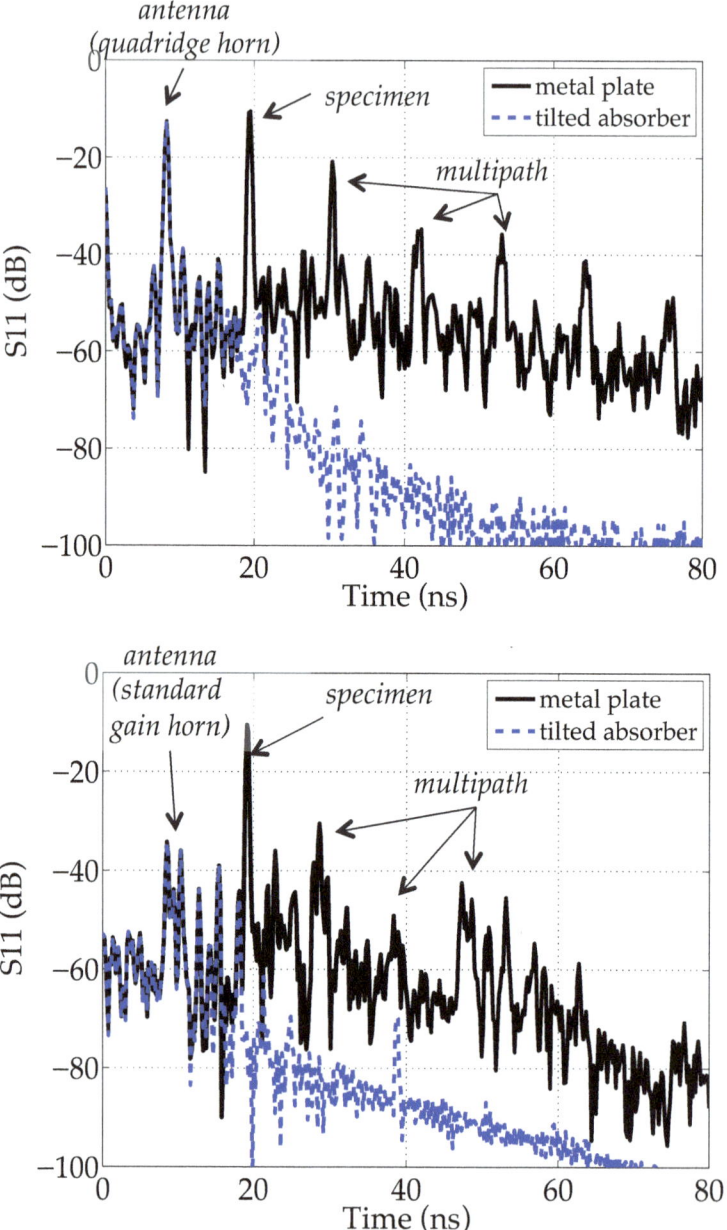

Figure 19   Measured focused beam reflection after Fourier transform
to time domain; quadridge horn feed (top) and X-band
standard gain horn feed (bottom).

Filtering measured data with a time-domain gate is very effective in 'cleaning up' the measured data. Gate width is typically specified in nanoseconds, and the choice of what width to use depends on the measurement frequency range and the characteristics of the measured specimen. Minimizing the width of the gate function also minimizes the ripple from undesired multi-path signals. However, if the gate width is less than the width of the desired signal, then this filtering will induce additional systematic errors in the result. In particular, care should be used for gating of resonant specimens (such as frequency selective surfaces, radomes, or metal-backed narrow-band absorbers) because they may 'ring' for a significant amount of time. A gate width that is used for a simple slab of material may be too narrow for these more resonant structures, even when the physical specimen thickness is the same.

3. INVERTING INTRINSIC PROPERTIES

MICROWAVE NETWORK ANALYSIS

Once calibrated and processed scattering parameters are obtained, then an inversion algorithm is applied. These inversion methods are derived by solving the boundary value problem of a plane wave interacting with a planar slab of material. In particular, the intrinsic properties can be determined by directly applying Maxwell's equations to the various boundaries in a given geometry and solving for the entire system at once. This approach is relatively straightforward for a single boundary, however as the complexity of a specimen geometry grows (e.g. multi-layer systems), property inversion becomes more and more complicated. As a result, boundary-value problems in microwave systems are often framed in terms of network analysis. Microwave network theory is a formalism that enables a more complicated transmission line problem to be solved by breaking it up into smaller pieces, which can then be recombined later by matrix multiplication to solve the entire problem.

In the most general case, a microwave network is a region of space having an arbitrary shape and having some number of waveguide or transmission line inputs and outputs. The inputs and outputs are also called ports. A simple free-space measurement of a single-layer dielectric or magnetic slab is an example of a two-port network, and we can evaluate the characteristics of that slab in terms of the fields at the two ports or faces of the material. Figure 20 shows a schematic representation of a two-port network with input voltages a_i and output voltages b_i. The input and output voltages can be related to each other by a matrix formulation known as a scattering matrix,

$$\begin{bmatrix} b_1 \\ b_2 \end{bmatrix} = \begin{bmatrix} S_{11} & S_{12} \\ S_{21} & S_{22} \end{bmatrix} \begin{bmatrix} a_1 \\ a_2 \end{bmatrix},$$ (3.1)

where S_{ij} are the elements of the scattering matrix, or "S-parameters". Another formulation that is more convenient for cascading multiple networks together is the R-matrix form,

$$\begin{bmatrix} b_1 \\ a_1 \end{bmatrix} = \begin{bmatrix} R_{11} & R_{12} \\ R_{21} & R_{22} \end{bmatrix} \begin{bmatrix} a_2 \\ b_2 \end{bmatrix}.$$ (3.2)

Thus it is necessary to be able to convert between these two forms. The following formulas can be used to convert back and forth between the R-matrix and S-matrix,

$$R = \frac{1}{S_{21}} \begin{bmatrix} S_{12}S_{21} - S_{11}S_{22} & S_{11} \\ -S_{22} & 1 \end{bmatrix}$$ (3.3)

$$S = \frac{1}{R_{22}} \begin{bmatrix} R_{12} & R_{11}R_{22} - R_{12}R_{21} \\ 1 & -R_{21} \end{bmatrix}.$$ (3.4)

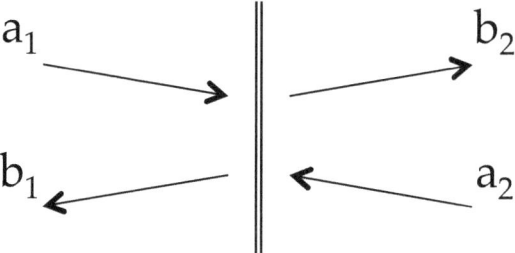

Figure 20   Simple 2-Port network showing input and output voltages.

In measurements of transmission and reflection from a material specimen, this network analysis formalism is used to derive the necessary relationships between the measured scattering parameters and the intrinsic properties of that specimen. For example, a simple slab of homogeneous and isotropic material can be defined in terms of a two-port network. However to derive a relationship between the network scattering parameters and the intrinsic material properties within the slab, it is first necessary to segment it into simpler parts. Specifically the slab can be defined in terms of three different cascaded two-port networks that represent the front surface, back surface, and region between the front and back, respectively. This system of cascaded networks is shown schematically in Figure 21. In the following, we calculate the R-matrix of each of the three networks by analyzing the voltages in terms of transmission, τ, or reflection, Γ coefficients; and in the case of region II, we analyze the R-matrix in terms of the wave propagation through the medium, T. The transmission coefficient is the ratio of the transmitted wave to the incident wave, and the reflection coefficient is the ratio of the reflected wave to the incident. The wave propagation factor, T, is the phase and amplitude change in the wave as it traverses through the thickness of the material specimen.

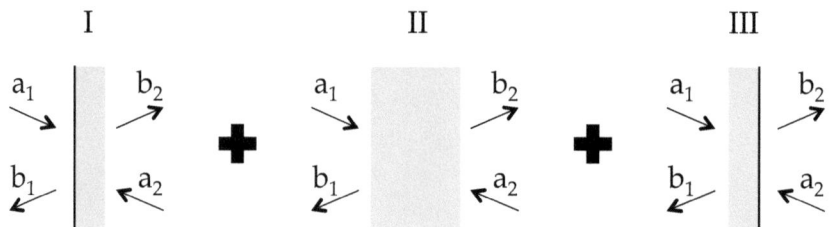

Figure 21   2-Port network showing input and output voltages for three different regions of a homogeneous slab specimen.

For this derivation, we first assume an incoming wave from either the left side travelling towards the right or from the right side travelling towards the left; and with a voltage, V_0. Region I in the figure above represents the left-most boundary between free-space, and the material specimen. So we can assign the network voltages of region I in terms of this input voltage:

Right travelling wave	*Left travelling wave*
$a_1 = V_0$	$a_1 = 0$
$b_1 = \Gamma^+ V_0$	$b_1 = \tau^- V_0$
$a_2 = 0$	$a_2 = V_0$
$b_2 = \tau^+ V_0$	$b_2 = \Gamma^- V_0$

Inserting these voltages into Equation (3.2) and recognizing that $\Gamma^+ = -\Gamma^-$, the R-matrix elements can then be determined:

$$R^I = \begin{bmatrix} \tau^- + \dfrac{\Gamma^{+2}}{\tau^+} & \dfrac{\Gamma^+}{\tau^+} \\[2ex] \dfrac{\Gamma^+}{\tau^+} & \dfrac{1}{\tau^+} \end{bmatrix} \tag{3.5}$$

At the boundary, the tangential E-fields of the incident and reflected waves must match that of the transmitted. Thus we can write equations that relate the transmission and reflection coefficients: $1 + \Gamma^+ = \tau^+$ and $1 + \Gamma^- = \tau^-$. Plugging these relationships into Equation (3.5) then provides the R-matrix of region I in terms of Γ^+,

$$R^I = \frac{1}{\tau}\begin{bmatrix} 1 & \Gamma \\ \Gamma & 1 \end{bmatrix} = \frac{1}{1+\Gamma}\begin{bmatrix} 1 & \Gamma \\ \Gamma & 1 \end{bmatrix}, \tag{3.6}$$

where the superscript '+' has been dropped for brevity. Additionally, the reflection coefficient can be expressed in terms of the intrinsic permittivity and permeability [39],

$$\Gamma = \frac{\sqrt{\dfrac{\mu}{\varepsilon}} - 1}{\sqrt{\dfrac{\mu}{\varepsilon}} + 1}, \tag{3.7}$$

where normal incidence transmission through the specimen is assumed.

Region II in Figure 21 represents the sample medium between the two air-material interfaces. For an incident wave with voltage of V_0 we can assign the network voltages of region II in terms of this input voltage:

Right travelling wave	Left travelling wave
$a_1 = V_0$	$a_1 = 0$
$b_1 = 0$	$b_1 = TV_0$
$a_2 = 0$	$a_2 = V_0$
$b_2 = TV_0$	$b_2 = 0$

where T is the propagation factor of the wave through region II, $T = e^{-ikt}$. $k = k_0\sqrt{\varepsilon_r \mu_r}$ is the wavenumber in the sample medium and t is the distance traveled between the two ports. For a simple slab with a normal-incidence illuminating wave, t is simply the thickness of the slab. Plugging these boundary conditions into Equation (3.2) then gives the R-matrix for region II,

$$R^{II} = \begin{bmatrix} T & 0 \\ 0 & \dfrac{1}{T} \end{bmatrix}. \tag{3.8}$$

Finally the R-matrix for the back face of the material slab, region III, follows a derivation similar to the front face. Because it is the reverse of the front face, it is essentially the same matrix, but with the reflection coefficients replaced by their negatives,

$$R^{III} = \frac{1}{1 - \Gamma} \begin{bmatrix} 1 & -\Gamma \\ -\Gamma & 1 \end{bmatrix}. \tag{3.9}$$

The equivalent R-matrix for all three regions put together is then calculated with a matrix multiplication,

$$R = R^{I} R^{II} R^{III} = \frac{1}{T(1 - \Gamma^2)} \begin{bmatrix} T^2 - \Gamma^2 & \Gamma - \Gamma T^2 \\ \Gamma T^2 - \Gamma & 1 - \Gamma^2 T^2 \end{bmatrix}. \tag{3.10}$$

Then with Equation (3.4), we can convert this R-matrix to the elements of the scattering matrix for the full material slab,

$$S = \frac{1}{1 - \Gamma^2 T^2} \begin{bmatrix} \Gamma(1 - T^2) & T(1 - \Gamma^2) \\ T(1 - \Gamma^2) & \Gamma(1 - T^2) \end{bmatrix}. \tag{3.11}$$

We now have a relationship between the scattering parameters that can be measured experimentally and the intrinsic permittivity and permeability of the material under test. In a few cases, Equation (3.11) can be written such that the permittivity and/or permeability are expressed as a direct function of the measured scattering parameters. More often however, an iterative root finding method must be used to determine the intrinsic properties.

The following paragraphs present a number of common inversion algorithms that are based on Equation (3.11) or other similar equations. This chapter thus provides a library of algorithms, which can be applied based on the needs of a given specimen measurement scenario. Note that in the following discussion, μ and ε are assumed to refer to the relative permeability and permittivity and the subscript r is dropped for brevity.

NICOLSON – ROSS – WEIR (NRW) ALGORITHM

The NRW algorithm is a well-known algorithm for inverting permittivity and permeability from the S_{11} and S_{21} network scattering parameters [40,41]. From a specimen's measured S_{11} and S_{21}, reflection, Γ, at the air/material boundary and transmission, T through the material are calculated,

$$\Gamma = X \pm \sqrt{X^2 - 1}, \quad \text{where} \quad X = \frac{S_{11}^2 - S_{21}^2 + 1}{2S_{11}} \tag{3.12}$$

$$T = \frac{S_{11} + S_{21} - \Gamma}{1 - (S_{11} + S_{21})\Gamma} \tag{3.13}$$

By defining a third parameter, Λ, such that,

$$\frac{1}{\Lambda^2} = -\left(\frac{1}{2\pi t} \ln T\right)^2, \tag{3.14}$$

then the permeability and permittivity can be solved for explicitly,

$$\mu = \frac{2\pi}{\Lambda k_0}\left(\frac{1 + \Gamma}{1 - \Gamma}\right) \tag{3.15}$$

$$\varepsilon = \frac{1}{\mu k_0^2}\left(\frac{4\pi^2}{\Lambda^2}\right) \tag{3.16}$$

where $k_0 = 2\pi/\lambda$ is the wavenumber in air. As Baker-Jarvis [42] points out, the NRW algorithm outlined above suffers from a numeric instability when the frequency corresponds to a multiple of one-half wavelength in the specimen. This instability is caused in part by the large phase uncertainties, and is more likely to happen in low-loss materials. Boughriet et al. [43] have noted that the instability arises from the computation of the factor, $(1 + \Gamma)/(1 - \Gamma)$. Therefore, if μ is already known, this instability can be circumvented since this factor affects the computation of ε only by its influence on μ.

Another difficulty with this algorithm stems from the fact that the logarithm of a complex number is multi-valued. In other words, there is a phase ambiguity in Equation (3.14) such that $\ln T = \ln T + i2\pi n$, where n is an integer. In most "nice" specimens, i.e. when the permittivity and permeability are moderate to small, and the sample electrical thickness is less than a half wavelength, the $n = 0$ solution is typically the correct one. However in some cases $n \neq 0$ and additional information is needed to identify the correct solution. For example, Baker-Jarvis et al [42], outline a procedure that compares measured and calculated group delays to identify which solution is correct. Alternatively, a priori knowledge of the expected range of permittivity and permeability values can also be used to determine the correct root.

ITERATIVE ALGORITHM – S11 OR S21

While the NRW algorithm is useful for many materials, there are cases where both S_{11} and S_{21} cannot be accurately measured. For example, in highly reflective materials, it may be that the S_{21} is so small as to be inaccurate. Conversely, low dielectric materials may have a very small S_{11}, so that a calculation based solely on S_{21} is preferable. For materials where the magnetic permeability is already known, an iterative calculation based on either S_{11} or S_{21} can be performed. All iterative algorithms begin with initial estimates at all frequencies for permittivity (and permeability when both parameters are unknown). The algorithm then refines the initial estimates with a numerical root-finding technique. There are often multiple solutions to these equations, and an initial estimate is necessary to select the proper root. Suitability of the initial estimates is determined by the stability of permittivity and permeability results, or by comparison with other measurement methods. If the initial estimates start the iterative calculation on the wrong root, the calculated results tend to have poor convergence and may jump to another root when plotted as a function of frequency. Improper solutions may also violate energy conservation by having negative loss.

For a two-port network, it is shown above that the reflection and transmission coefficients defined above are related to the scattering parameters, S_{11} and S_{21} by [42],

$$S_{11} = \frac{\Gamma(1 - T^2)}{1 - \Gamma^2 T^2} \quad \text{and} \quad S_{21} = \frac{T(1 - \Gamma^2)}{1 - \Gamma^2 T^2} \tag{3.17}$$

where S_{11} and S_{21} are the calibrated S-parameters. These equations also assume that a phase correction was applied for the displacement of the transmission path by the specimen. The general relations for the reflection and transmission coefficients at arbitrary incidence angle are,

$$\Gamma_{TE} = \frac{\mu \cos \theta - \sqrt{\mu \varepsilon - \sin^2 \theta}}{\mu \cos \theta + \sqrt{\mu \varepsilon - \sin^2 \theta}} \tag{3.18}$$

$$\Gamma_{TM} = \frac{\sqrt{\mu \varepsilon - \sin^2 \theta} - \varepsilon \cos \theta}{\sqrt{\mu \varepsilon - \sin^2 \theta} + \varepsilon \cos \theta} \tag{3.19}$$

$$T = e^{-ik_0 t \sqrt{\mu \varepsilon - \sin^2 \theta}} \tag{3.20}$$

where t is the thickness of the specimen and θ is the angle between the direction of propagation and the specimen normal. Equations (3.18) and (3.19) are derived by the matching boundary conditions at the specimen/air interface. Equation (3.20) is the propagation through a specimen of thickness t and at an angle θ. Note that the angle inside the specimen is different than the incidence angle. However the angle within the specimen was eliminated from the above equations by using Snell's law along with a geometric identity. Typical measurements orient the specimen at normal incidence ($\theta = 0$), and the above equations reduce to,

$$\Gamma = \frac{\sqrt{\frac{\mu}{\varepsilon}} - 1}{\sqrt{\frac{\mu}{\varepsilon}} + 1} \quad \text{and} \quad T = e^{-ik_0 t \sqrt{\mu \varepsilon}} \tag{3.21}$$

Either of the equations in (3.17) along with the reflection or transmission coefficient as a function of frequency can then be solved via Newton's method using ε or μ as the unknown variable. Newton's method is recommended since it has the ability to solve for complex valued roots [44]. It also requires the evaluation of the first derivative of the function. The reformulated functions and their derivatives for solving S_{11} and S_{21} are as follows,

S_{11}:

$$f = 0 = (1 - \Gamma^2 T^2)S_{11} - \Gamma(1 - T^2) \qquad (3.22)$$

$$f' = [T^2(1 - 2S_{11}\Gamma) - 1]\Gamma' - 2T\Gamma(1 - S_{11}\Gamma)T' \qquad (3.23)$$

S_{21}:

$$g = 0 = (1 - \Gamma^2 T^2)S_{21} - T(1 - \Gamma^2) \qquad (3.24)$$

$$g' = [\Gamma^2(1 - 2S_{21}T) - 1]T' + 2T\Gamma(1 - S_{21}T)\Gamma' \qquad (3.25)$$

While this iterative technique will work for any value of μ, it is usually only applied to dielectric materials, where $\mu = 1$ (since a separate measurement of μ would otherwise be required). The derivatives of the reflection and transmission coefficients with respect to permittivity are,

$$\frac{\partial \Gamma_{TE}}{\partial \varepsilon} = \frac{-\mu^2 \cos \theta}{\sqrt{\mu\varepsilon - \sin^2 \theta} \left(\mu \cos \theta + \sqrt{\mu\varepsilon - \sin^2 \theta}\right)^2} \qquad (3.26)$$

$$\frac{\partial \Gamma_{TM}}{\partial \varepsilon} = \frac{\cos \theta \left(\dfrac{\mu\varepsilon}{\sqrt{\mu\varepsilon - \sin^2 \theta}} - 2\sqrt{\mu\varepsilon - \sin^2 \theta}\right)}{\left(\varepsilon \cos \theta + \sqrt{\mu\varepsilon - \sin^2 \theta}\right)^2} \qquad (3.27)$$

$$\frac{\partial T}{\partial \varepsilon} = -\frac{ik_0\mu t}{2\sqrt{\mu\varepsilon - \sin^2 \theta}}T \qquad (3.28)$$

For S_{21} measurements, Newton's method solves these equations by iteratively calculating $\varepsilon_{n+1} = \varepsilon_n - g/g'$ until $\varepsilon_{n+1} - \varepsilon_n$ is sufficiently small.

Since iterative methods rely on starting guess values, it may be beneficial in some cases to iterate serially -- one frequency at a time. For a material that is not very dispersive and the permittivity and permeability are stable across the measurement bandwidth, the same starting guess value can work for all of the frequency dependent data. However, for a material that is dispersive (i.e. has a frequency dependent permittivity or permeability), conducting the iterative procedure in series means that a guess value is only needed for the first frequency point, and the converged solution for that frequency then becomes the guess value for the next frequency point. Otherwise, a frequency dependent set of guess values is needed to get good convergence for the whole bandwidth, which could be especially difficult to estimate for a material that is undergoing a relaxation within the measurement bandwidth. Instead, it is easier to start at one end or the other of the band where the material permittivity or permeability may be better known. Often it is easiest to start at the lowest frequency and work upwards, since the material is electrically thinner at lower frequencies, and the multiple solution branches are spaced further apart.

ITERATIVE ALGORITHM – S11 AND S21

An iterative algorithm to solve for μ and ε simultaneously can also be implemented based on measurements of both S_{11} and S_{21}. Because both S_{11} and S_{21} equations have to be solved, Newton's iteration for a system of equations is used. In this case of more than one variable to be solved, the algorithm is most easily expressed in matrix form [45] where the following linear system is solved,

$$
\begin{bmatrix}
\dfrac{\partial f}{\partial \varepsilon} & \dfrac{\partial f}{\partial \mu} \\[2mm]
\dfrac{\partial g}{\partial \varepsilon} & \dfrac{\partial g}{\partial \mu}
\end{bmatrix}
\cdot
\begin{bmatrix} \Delta\varepsilon \\ \Delta\mu \end{bmatrix}
=
\begin{bmatrix} f \\ g \end{bmatrix}
\tag{3.29}
$$

and where f and g and their derivatives are defined previously by Equations (3.22) thru (3.25). Since μ is no longer assumed fixed, this algorithm also requires the derivatives of the reflection and transmission coefficients with respect to μ,

$$
\frac{\partial \Gamma_{TE}}{\partial \mu} = \frac{\cos\theta \left(-\dfrac{\mu\varepsilon}{\sqrt{\mu\varepsilon - \sin^2\theta}} + 2\sqrt{\mu\varepsilon - \sin^2\theta} \right)}{\left(\mu\cos\theta + \sqrt{\mu\varepsilon - \sin^2\theta} \right)^2}
\tag{3.30}
$$

$$\frac{\partial \Gamma_{TM}}{\partial \mu} = \frac{\varepsilon^2 \cos \theta}{\sqrt{\mu \varepsilon - \sin^2 \theta} \left(\varepsilon \cos \theta + \sqrt{\mu \varepsilon - \sin^2 \theta} \right)^2} \qquad (3.31)$$

$$\frac{\partial T}{\partial \mu} = -\frac{i k_0 \varepsilon t}{2 \sqrt{\mu \varepsilon - \sin^2 \theta}} T \qquad (3.32)$$

The two-by-two matrix of Equation (3.29) is the Jacobian of the system of two complex equations. In matrix notation the discretized system of equations can be represented as follows

$$J \Delta X = Y \qquad (3.33)$$

where the vector X contains the estimated values of permittivity and permeability, and the vector Y contains the zero valued functions f and g given above. Solving for ΔX gives $\Delta X = J^{-1} Y$. Similar to the single equation iteration algorithm discussed above, to solve for permittivity and permeability, this algorithm starts with an initial estimate of X and calculates ΔX. The functional iteration procedure is then $X^{new} = X^{old} - J^{-1} Y$, which is repeated until $X^{new} \approx X^{old}$, at which point the converged values of permittivity and permeability have been found. This iterative algorithm should give the same results as the more direct NRW method so it is of limited usefulness. It also suffers from the same $\lambda/2$ wavelength instability as the NRW algorithm.

ITERATIVE ALGORITHM – SHORTED S$_{11}$

The iterative algorithm for S_{11} outlined above assumes that there is no short (conductive metal plate) behind the specimen. However it is also possible to calculate permittivity when there is a conductive plate (electrical short) behind the specimen. This may be useful for materials that are applied directly to a ground plane and where separating the specimen from the ground plane is difficult. This method was originally developed by Roberts and von Hippel [46], and is reviewed by Baker-Jarvis [42]. When the specimen is flush against the short, the permittivity and permeability are related to S_{11} by,

$$S_{11} = \frac{\mu \tanh(i\sqrt{\mu\varepsilon}k_o t) - \sqrt{\mu\varepsilon}}{\mu \tanh(i\sqrt{\mu\varepsilon}k_o t) + \sqrt{\mu\varepsilon}} \qquad (3.34)$$

Similar to the previous iteration methods, Equation (3.34) is reformulated so that Newton's root finding method can be applied to iteratively solve for ε. In particular the iterated function and its derivative are as follows,

$$f(\varepsilon) = 0 = \left(\tanh\left(i\sqrt{\varepsilon}k_o t\right) + \sqrt{\varepsilon}\right)S_{11} - \left(\tanh\left(i\sqrt{\varepsilon}k_o t\right) - \sqrt{\varepsilon}\right) \qquad (3.35)$$

$$f'(\varepsilon) = \frac{k_0}{2\sqrt{\varepsilon}}\left[\begin{array}{l}\left(k_0 \tanh^2\left(i\sqrt{\varepsilon}k_o t\right) + k_0 t + i\right)S_{11} \\ -k_0 \tanh^2\left(i\sqrt{\varepsilon}k_o t\right) + k_0 t + i\end{array}\right] \qquad (3.36)$$

ITERATIVE ALGORITHM – SHORTED S_{11} AND S_{21}

The shorted S_{11} iteration equations can be combined with the equations for S_{21} so that magnetic specimens can be measured. As in the non-shorted S_{11} and S_{21} iteration, this algorithm uses Newton's method for a system of equations as outlined in Equation (3.29). In this case, the iterated function, f, is obtained from Equation (3.34), and the second iterated function, g, is obtained from Equations (3.17),

$$f(\varepsilon,\mu) = \left(\mu \tanh\left(i\sqrt{\mu\varepsilon}k_o t\right) + \sqrt{\mu\varepsilon}\right)S_{11} - \left(\mu \tanh\left(i\sqrt{\mu\varepsilon}k_o t\right) - \sqrt{\mu\varepsilon}\right) \qquad (3.37)$$

$$g(\varepsilon,\mu) = (1 - \Gamma^2 T^2)S_{21} - T(1 - \Gamma^2) \qquad (3.38)$$

The derivatives of these two functions for solving via Newton's algorithm are,

$$\frac{\partial f}{\partial \varepsilon} = \frac{k_0^2 \mu^2 t}{2\sqrt{\mu\varepsilon}}\left[\begin{array}{l}\left(\tanh^2\left(i\sqrt{\mu\varepsilon}k_o t\right) - 1 + \dfrac{i}{\mu k_0 t}\right)S_{11} \\ -\tanh^2\left(i\sqrt{\mu\varepsilon}k_o t\right) + 1 + \dfrac{i}{\mu k_0 t}\end{array}\right] \qquad (3.39)$$

$$\frac{\partial f}{\partial \mu} = \frac{k_0}{2\sqrt{\mu\varepsilon}} \left[\begin{array}{c} \left(\dfrac{\mu\varepsilon t k_0 \tanh^2(i\sqrt{\mu\varepsilon}k_0 t)}{-2i\sqrt{\mu\varepsilon}\tanh(i\sqrt{\mu\varepsilon}k_0 t) - k_0\mu\varepsilon t + i\varepsilon} \right) S_{11} \\[2mm] -\mu\varepsilon t k_0 \tanh^2(i\sqrt{\mu\varepsilon}k_0 t) \\[1mm] +2i\sqrt{\mu\varepsilon}\tanh(i\sqrt{\mu\varepsilon}k_0 t) + k_0\mu\varepsilon t + i\varepsilon \end{array} \right] \quad (3.40)$$

$$\frac{\partial g}{\partial \varepsilon} = [\Gamma^2(1 - 2S_{21}T) - 1]\frac{\partial T}{\partial \varepsilon} + 2T\Gamma(1 - S_{21}T)\frac{\partial \Gamma}{\partial \varepsilon} \qquad (3.41)$$

$$\frac{\partial g}{\partial \mu} = [\Gamma^2(1 - 2S_{21}T) - 1]\frac{\partial T}{\partial \mu} + 2T\Gamma(1 - S_{21}T)\frac{\partial \Gamma}{\partial \mu} \qquad (3.42)$$

These equations along with the derivatives for Γ and T given above are solved iteratively via Equation (3.29).

ITERATIVE ALGORITHM – 4-PARAMETER

A significant disadvantage of the above algorithms that utilize S_{11} is that they can have significant measurement uncertainties. These errors arise from warped specimen geometries, where slight or not-so-slight curvature occurs due to material flexibility or internal material stresses. Specimen curvature results in a net position offset of the specimen from the reflection calibration plane, causing a phase error in the measured reflection coefficient. Instead of just two scattering parameters, the four-parameter method uses all four scattering parameters (S_{11}, S_{22}, S_{21}, S_{12}) to determine permittivity and permeability. This method has the advantage of not needing a precise placement of the specimen relative to the calibration reference plane. In particular, the phase error associated with specimen placement is eliminated, resulting in greater accuracy for the permittivity and permeability calculations.

At first glance, using forward and backward scattering coefficients over-specifies the inversion problem since only two complex intrinsic properties are calculated. For homogeneous specimens, the transmission and reflection amplitudes as well as the transmission phases should all be the same. Only the reflection phases should differ, depending on specimen position. Thus the four-parameter inversion includes additional information that can determine specimen position without a physical distance measurement. These additional parameters are used to algebraically eliminate the specimen position dependence.

The Four-parameter inversion method for a focused beam system is similar to a procedure often used in coaxial airline fixtures [45]. The equations relating the scattering parameters to specimen thickness, permittivity and permeability are as follows,

$$S_{11}^{cal} S_{22}^{cal} e^{-2ik_0(t_s - t_m)} - S_{21}^{cal} S_{12}^{cal} e^{-2ik_0(t_s)} = \frac{\Gamma^2 - T^2}{1 - \Gamma^2 T^2} \qquad (3.43)$$

$$e^{-ik_0 t_s} \frac{S_{21}^{cal} + S_{12}^{cal}}{2} = \frac{T(1 - \Gamma^2)}{1 - \Gamma^2 T^2} \qquad (3.44)$$

where $\Gamma = (\mu - \sqrt{\mu\varepsilon})/(\mu + \sqrt{\mu\varepsilon})$ and $T = e^{-ik_0 t_s \sqrt{\mu\varepsilon}}$. The 'cal' superscript designates the use of already calibrated scattering parameters, and the algorithm must also account for the transmission line displaced with the metal calibration plate by including the calibration plate thickness, t_m as well as the specimen thickness, t_s. This algorithm also requires the use of the four-parameter calibration method described above, where S_{11} and S_{22} response calibration factors are measured simultaneously without moving the metal calibration standard.

These equations are then solved to invert permittivity and permeability using the system of equations defined by Equation (3.29). The zero-valued functions whose roots must be obtained are

$$f = (1 - \Gamma^2 T^2)\left(S_{11}^{cal} S_{22}^{cal} e^{-2ik_0(t_s - t_m)} - S_{21}^{cal} S_{12}^{cal} e^{-2ik_0(t_s)}\right) \\ - (\Gamma^2 - T^2) \qquad (3.45)$$

$$g = (1 - \Gamma^2 T^2)\left(e^{-ik_0 t_s} \frac{S_{21}^{cal} + S_{12}^{cal}}{2}\right) - T(1 - \Gamma^2) \qquad (3.46)$$

INVERTING SHEET IMPEDANCE

It is sometimes convenient to represent a material as a shunt impedance. Certain classical absorber materials such as the Salisbury screen and Jaumann absorber [47] are constructed using resistive sheets or "R-cards," where the shunt impedance is expressed in units of ohms/square [48]. In this case the sheet impedance is directly related to the scattering parameters, and can be derived using the same network analysis formalism shown at the beginning of this chapter.

The shunt resistance of the sheet can be treated as a simple two-port network. When an incident wave, with a voltage, V_0, interacts with the sheet, the network voltages are as follows:

Right travelling wave	Left travelling wave
$a_1 = V_0$	$a_1 = 0$
$b_1 = \Gamma^+ V_0$	$b_1 = \tau^- V_0$
$a_2 = 0$	$a_2 = V_0$
$b_2 = \tau^+ V_0$	$b_2 = \Gamma^- V_0$

Unlike the interface between air and a dielectric material discussed previously, the discontinuity represented by the sheet is symmetrical so that $\Gamma^+ = \Gamma^-$ and $\tau^+ = \tau^-$. Additionally the tangential E-fields should match so the relationship, $1 + \Gamma = \tau$ applies. Finally, with straightforward circuit analysis [48], the reflection coefficient from a shunt impedance Z, in a transmission line with a characteristic impedance Z_0 is given by $\Gamma = (Z - Z_0)/(Z + Z_0)$. These relationships then lead to the following R-matrix for an R-card,

$$R = \frac{1}{2Z_s}\begin{bmatrix} 2Z_s - Z_0 & -Z_0 \\ Z_0 & 2Z_s + Z_0 \end{bmatrix}. \tag{3.47}$$

With Equation (3.4), this R-matrix can then be converted into scattering parameters. For transmission,

$$Z_s = \frac{Z_0 S_{21}}{2(1 - S_{21})} \tag{3.48}$$

and for reflection,

$$Z_s = \frac{-Z_0(1 + S_{11})}{2 S_{11}} \tag{3.49}$$

where Z_0 is the bulk impedance of free space (377 ohms). Unlike the permittivity and permeability inversion, sheet impedance is calculated directly from the scattering parameters and an iterative inversion method is not needed.

Sometimes an R-card material is mounted onto a thick dielectric substrate. The effect of an additional substrate is to add additional capacitance to the complex impedance, resulting in a more negative imaginary impedance. When this is the case, it may be desirable to extract the impedance of just the R-card while excluding the effect of the supporting substrate. Using the cascade matrix theory, the R-matrix of the sheet is multiplied with the R-matrix of the dielectric slab to obtain an R-matrix, (and corresponding S-matrix) for the multilayer stack. If the substrate permittivity, ε, and thickness, t are accurately known, then the sheet impedance of just the R-card by itself can be calculated from the transmission scattering parameter by,

$$Z_s = S_{21} Z_0 \frac{\Gamma^2 \tau^2 - 1 + \Gamma(T^2 - 1)}{2(\Gamma^2 - 1) - 2S_{21}(\Gamma^2 T^2 - 1)}, \tag{3.50}$$

where $\Gamma = (\mu - \sqrt{\mu\varepsilon})/(\mu + \sqrt{\mu\varepsilon})$ and $T = e^{-ik_0 t\sqrt{\mu\varepsilon}}$ are of the substrate measured by itself (without a conductive layer). Thus this two-layer inversion requires that a specimen of the substrate be measured first without any R-card.

4. MEASUREMENT EXAMPLES

Representative examples of experimental material measurements are presented in this section to illustrate some of the procedures described above. The intent is to provide some insight into the characteristics of the data typically measured with a free space fixture. With the relative complexity of these measurement fixtures combined with the wide variety of different specimen characteristics that are possible, it is easy to acquire data but sometimes difficult to obtain *accurate* data. Some of the examples presented here represent very challenging measurements that can stress the accuracy of these measurement systems. Thus these examples illustrate some of the measurement issues that can occur and strategies for overcoming them.

A homogeneous, non-magnetic specimen provides a first check of the method's accuracy. A widely known and well-characterized dielectric material commonly used in microwave measurements is cross-linked polystyrene (trade name is Rexolite). It has a real dielectric permittivity that is approximately 2.54 at microwave frequencies. It has very little frequency dispersion so the imaginary part of the permittivity is very small – smaller than can be typically measured using transmission and reflection techniques. Figure 22 shows the inverted real and imaginary permittivity determined for a 3.3 mm thick Rexolite specimen measured in a focused beam system. For comparison, real and imaginary permittivity inverted with different inversion methods on the same measured data are plotted. In all four cases, a fixed time-domain window width (0.75 ns) was applied after calibration. The four-parameter method shows excellent agreement with the known permittivity and dielectric loss of Rexolite. The two-parameter and one-parameter, reflection (S_{11}) methods, however, show significant deviations in the form of unexpected frequency dispersion in the real and imaginary permittivity. The sign convention used for these data is $\varepsilon^* = \varepsilon' - i\varepsilon''$, so a negative imaginary permittivity indicates gain. Material gain is not physically possible, so these dispersive data (two-parameter and S_{11} inversions) are erroneous. On the other hand, the 4-parameter and S_{21} imaginary data are very close to zero and similarly, any excursion into negative imaginary permittivity is a result of measurement uncertainty.

The common factor between the two-parameter and S_{11} inversions causing erroneous results is the S_{11} data. These data are inaccurate because the reference plane defined by the metal calibration plate and the front face of the specimen are not in exactly the same position. In particular the specimen is not completely flat, and is slightly bowed. Even with a small amount of bowing (e.g. several hundred microns displacement between specimen and reference), significant reflection phase errors can occur, and these errors increase as frequency increases, i.e. as the displacement increases in electrical size. One method that has been used in the past to correct for this error is to manually fit the data with a reflection phase correction that eliminates non-physical imaginary permittivities and permeabilities. However this is an unsatisfactory solution that is at best semi-empirical.

On the other hand, the four-parameter inversion accounts for specimen/reference plane differences by measuring reflection in both directions. This limits the phase error to the uncertainty in the calibration plate thickness. Confirmation of this phase error phenomena can be seen in the one-parameter, transmission (S_{21}) data shown in Figure 22. These data show accuracies similar to the four-parameter results. The S_{21} inversion does not include reflection, so the errors associated with reflection reference plane displacement are non-existent. This is why S_{21} inversion is typically preferred for non-magnetic specimens. Unfortunately it is not an option when magnetic specimens are measured, and that is the primary advantage of the four-parameter method.

The data of Figure 22 show that the four-parameter inversion minimized errors from small, sub-millimeter specimen displacements. Figure 23 shows the measured real permittivity and real permeability of the same Rexolite specimen with the four-parameter inversion when displaced by large distances. These plots each show five curves where the specimen was displaced in 6.35 mm increments. Even at the largest displacement of 25.4 mm from the calibration plate reference plane, the measurement error has only increased slightly – still less than a few percent. This slight error increase is likely due to the finite focal depth of the Gaussian beam.

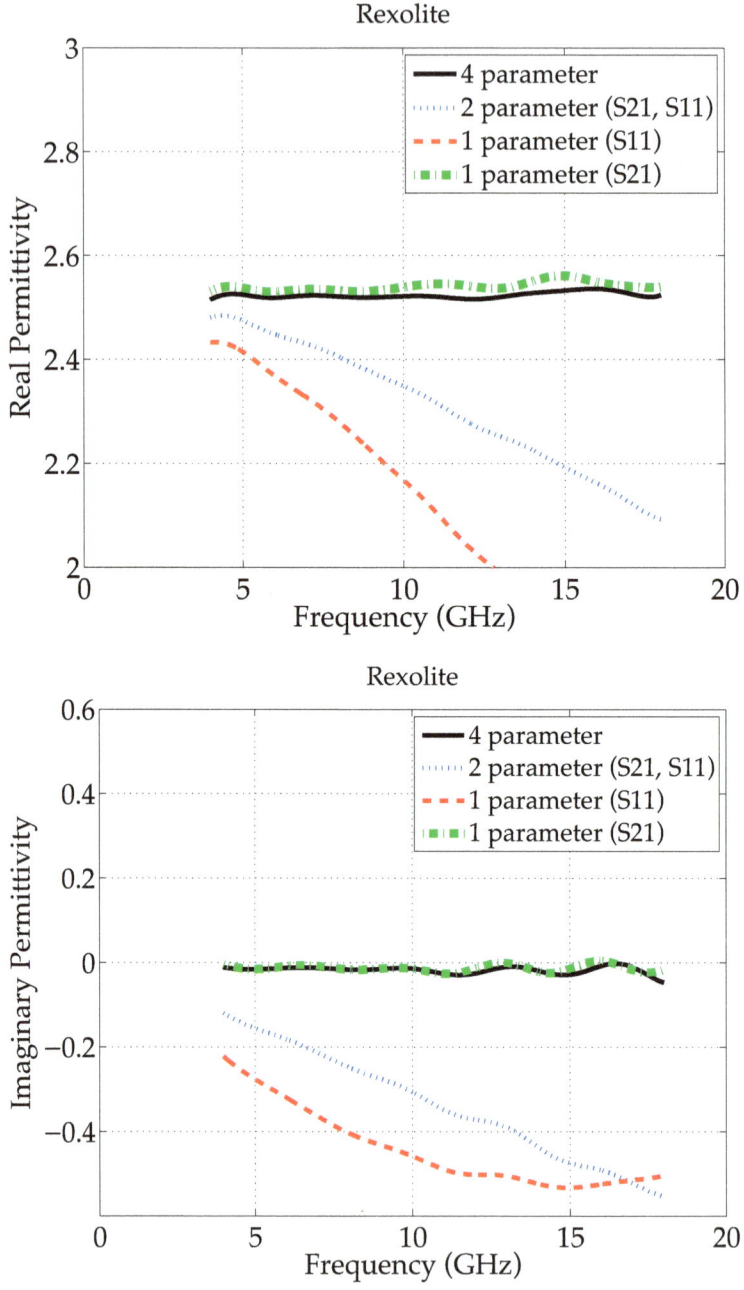

Figure 22   Real and imaginary permittivity of Rexolite (polystyrene) measured with focused beam and inverted from various network parameters.

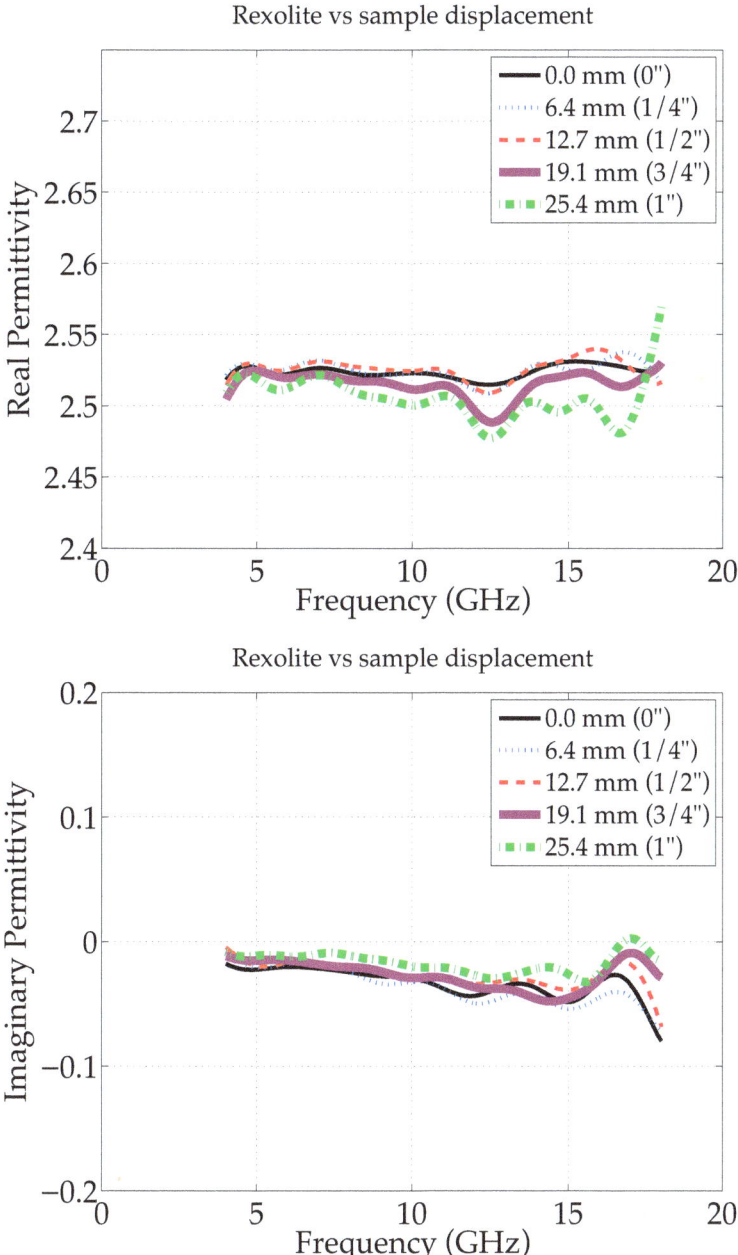

Figure 23   **The real permittivity and permeability of Rexolite (crosslinked polystyrene) measured with four-parameter inversion as a function of specimen displacement.**

As an alternative method for minimizing the reflection phase error, a mechanical or non-contact laser micrometer fixture can be incorporated into the focused beam system and used to physically measure the difference between the specimen and the calibration plate location. The additional apparatus however may not be as convenient as the four-parameter method since it adds more steps to the measurement procedure, while measuring scattering parameters in both directions is easily automated. Nevertheless, the four-parameter method assumes single-layer, reciprocal specimens. So if multilayer specimens are measured and reflection phase is important, then micrometer measurements will be necessary for accurately obtaining reflection phase.

Another example measurement of a homogeneous dielectric material is shown in Figure 24. The data shown at the top of Figure 24 is the measured real permittivity for a very thin polyimide sheet with a thickness of only 75 microns (0.003 inches). At 10 GHz, this specimen thickness is approximately $1/400^{th}$ of a wavelength, so it exhibits a very small perturbation on the transmitted wave in a focused beam system. Two curves are shown in the upper plot of Figure 24, each corresponding to a different calibration. In this case, only the transmission coefficient is used to invert permittivity, so the calibration depends primarily on a clear site measurement (S_{21} with no specimen). Both curves are based on the same specimen measurement, however the first calibration was conducted just a few minutes before the specimen measurement and the second calibration occurred a few minutes after the specimen measurement. The total time between the two calibrations was only 9 minutes, and the specimen measurement occurred in the time between the two.

The bottom plot in Figure 24 shows the phase difference between the two calibration measurements. In low-loss dielectric materials, the phase delay that is measured is dominated by the real part of the dielectric permittivity and its effect on the speed of light through the specimen. So the small difference observed between the phases of the two calibrations is responsible for the difference in measured permittivity shown in the top plot of Figure 24, and the two curves shown indicate that that there was a phase drift occurring over the course of the measurement. If that phase drift is linear and the measurement was made equidistant in time from the two calibrations, then a simple average of these two curves will correct for that phase drift. The expected permittivity for this polyimide measurement is approximately $\varepsilon = 3.5$, which roughly corresponds to the average value between the two measured curves. Because this specimen is so thin and the measured phase is so small to begin with, it is susceptible to phase drifts of even a tenth of a degree or less.

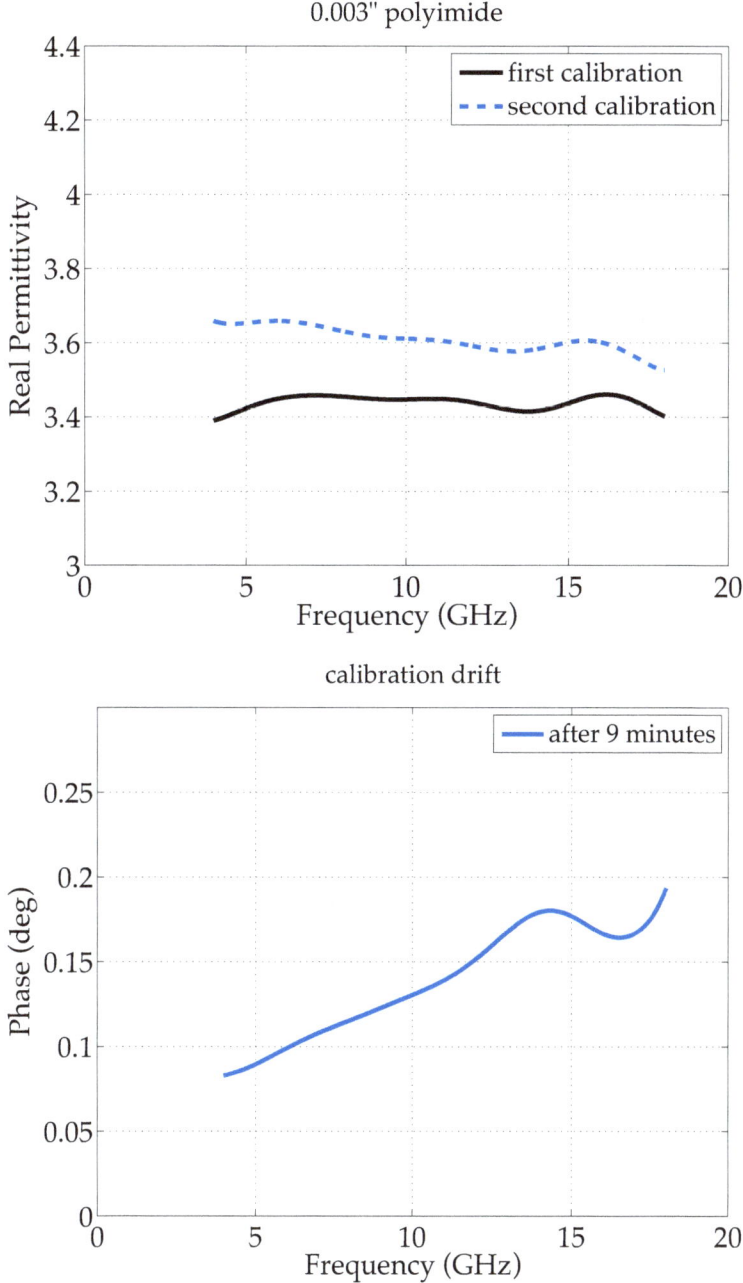

Figure 24   top: Measured permittivity for a thin (0.003″ / 76 micron)
polyimide sheet for two different calibrations – before
and after the specimen measurement; bottom: relative
phase drift during 9 minute time between the two
calibration measurements.

Modern network analyzers have highly stable sources, so phase drift from the internal electronics is generally very small. In a focused beam system, the occurrence of phase drift is more likely to be due to ambient temperature changes, i.e. the stability of the temperature of the laboratory in which the system resides. In particular, the microwave cables that connect the network analyzer to the feed antennas are coaxial transmission lines with the center conductor separated from the outer conductor by a Teflon-based dielectric. Teflon typically undergoes a material phase transition near room temperature that is reflected in the thermal expansion coefficient and results in a rapid dimensional change as a function of temperature. This phenomenon is sometimes referred to as the "Teflon knee" for the shape of the phase versus temperature curve that this dimensional change causes.

Thus ambient temperature changes of even a fraction of a degree can induce significant phase errors in the cables, including the phase drift observed in Figure 24. In thicker specimens, the phase delay induced by the specimen is much larger and this small phase drift is not a significant factor. However in the 76 micron thick specimens of Figure 24, the material induced phase delay is small so that the cable caused phase drift is much more noticeable.

Consequently is it important to monitor for phase drift error, and to correct for it when it occurs. For example, maintaining a steady temperature in the laboratory is important, as is calibrating often enough so that phase drift between calibrations is minimized. In cases such as Figure 24, where even a very small phase drift dominates, some laboratories have resorted to actively cooling their microwave cables. This can be as simple as wrapping the microwave cable with copper tubing that circulates chilled water. Cooling the cables to temperatures away from the "Teflon knee" temperature and where the thermal expansion coefficient is relatively low reduces variations in phase to an almost negligible level, even over longer (one to three hour) measurement intervals.

Moving beyond simple dielectric specimens, an example inversion for a magnetic material is shown in Figure 25 (dielectric permittivity) and Figure 26 (magnetic permeability). These results are for a commercial magnetic absorber material made from carbonyl iron powder mixed with an elastomer, and four different inversions are compared. The NRW and iterative two-parameter inversions are shown in the dashed red and continuous blue curves respectively. Both methods give almost identical results, which is because they are based on the same scattering parameters. Because the magnetic specimen was flexible, both of these inversions have significant errors due to the displacement of the specimen relative to the reference plane defined by the metal calibration plate.

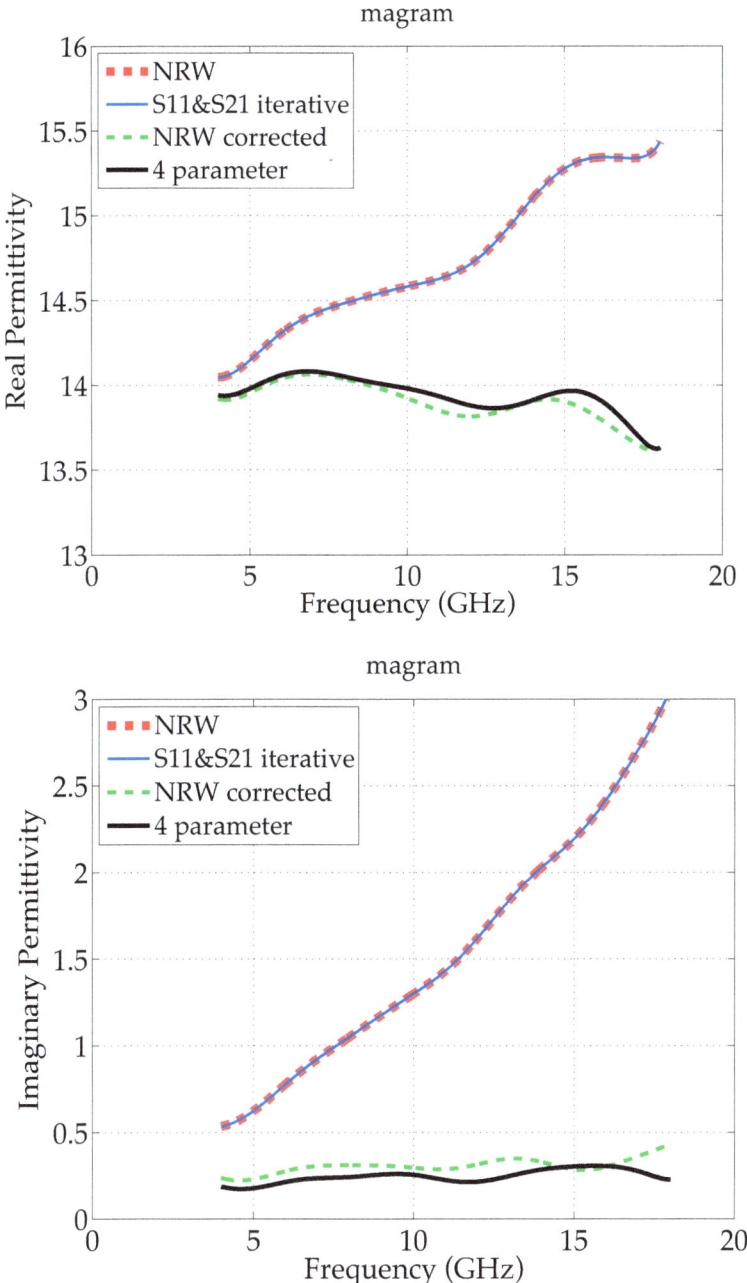

Figure 25   Real and imaginary permittivity for a commercial
magnetic absorber using various inversion algorithms.

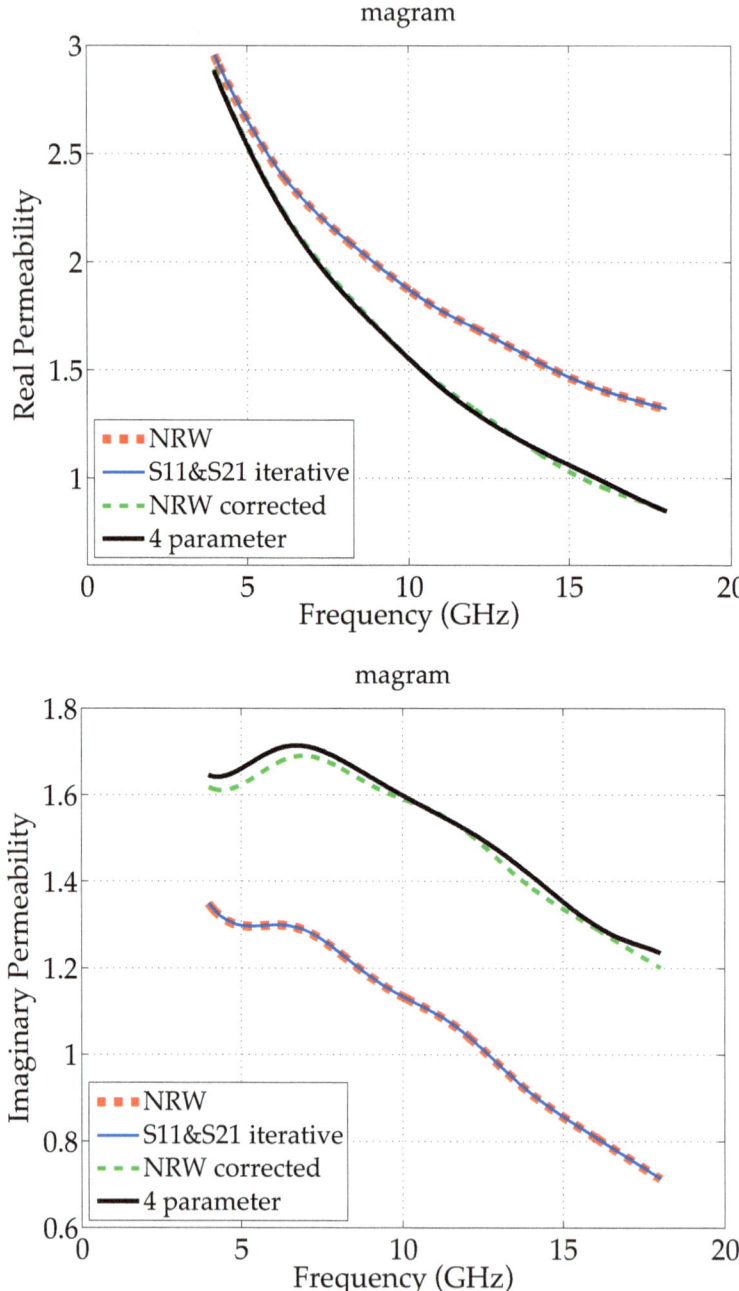

Figure 26   Real and imaginary permeability for a commercial
magnetic absorber using various inversion algorithms.

Also shown on these plots are curves from a 'corrected' NRW method and the four-parameter inversion method in dashed green and continuous black lines respectively. The corrected NRW method applied a phase offset corresponding to a net 0.32 mm displacement of the specimen from the calibration plate. This displacement was measured with a mechanical micrometer fixture similar to that discussed above. The micrometer was positioned in the fixture after the metal calibration plate was measured to obtain a reference location for the metal plate. The metal plate was then removed and the specimen was inserted and the micrometer then adjusted to determine the mechanical offset at a position corresponding to the center of the illuminating beam. The micrometer was then removed so that it would not interfere with the beam during the specimen measurement.

The agreement between this corrected NRW and the four-parameter method further demonstrates the importance of accounting for the reflection phase due to specimen displacement. The net effect of the phase error in the un-corrected inversions was a frequency dependent offset of both the permittivity and permeability. The amount of offset increased with frequency because the electrical size of the 0.32 mm specimen displacement also increased with frequency. One final observation: unlike the other two inversion methods, the corrected NRW and the four-parameter curves do not exactly overlay. This is reasonable because they use different sets of scattering parameters.

A second set of inverted properties for a much more challenging material specimen are shown in Figure 27. This material is an anisotropic composite of magnetic inclusions in a polymer matrix, designed to work as an electromagnetic interference (EMI) absorber in consumer electronics devices. The high aspect ratio of the magnetic particles gives them a large dipole moment resulting in extreme dielectric permittivity. Because the magnetic relaxation is primarily at frequencies in the VHF and UHF bands (30 MHz to 3 GHz), the focused beam data of Figure 27 show just the high frequency tail of the permeability relaxation.

The inverted data of Figure 27 were obtained using the four-parameter inversion method, which eliminates phase errors in the reflection coefficients. While using the more traditional NRW or S11 and S21 iteration methods results in significant errors in the magnetic specimens of Figure 25 and Figure 26, the more extreme permittivity of the EMI specimen results in a dramatically increased sensitivity to phase errors. Thus attempts to use the NRW or two-parameter iteration methods do not even converge and the four-parameter method is the only option in for this specimen.

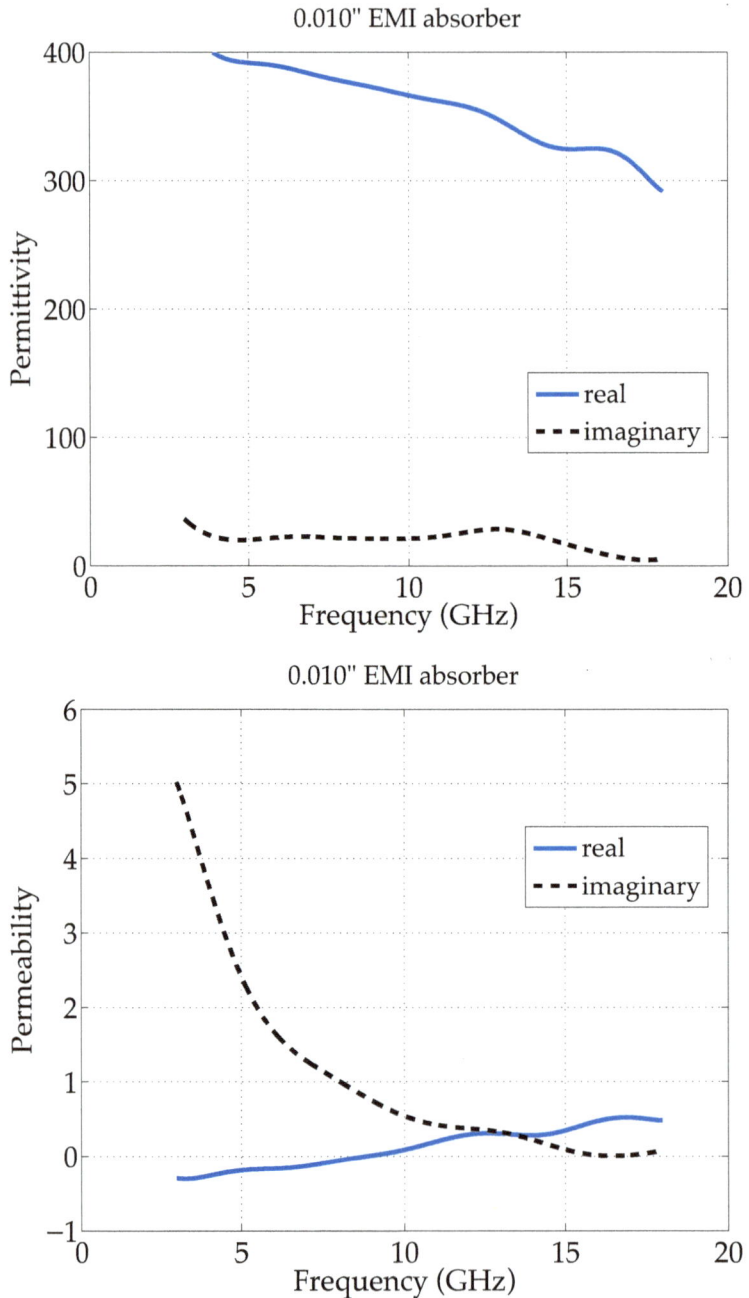

Figure 27   Measured permittivity and permeability for a high-index,
0.25 mm (0.010″) thick EMI absorber material.

Calculating material inversions from scattering parameter data is often difficult since the relationships between the scattering parameters and the intrinsic properties are not monotonic, meaning that there are frequently multiple solutions. When the iterative methods are used, the initial guess values must be reasonably close or the solver will run the risk of converging onto an incorrect solution. Obtaining reasonable results usually requires some foreknowledge of the expected range for the intrinsic properties. In difficult-to-measure specimens such as that of Figure 27, these difficulties are exacerbated. Thus it is sometimes advantageous to supplement focused beam measurements with other techniques, such as cavity or transmission line methods in order to have better bounds on the expected properties for the inversion.

5.   MEASUREMENT UNCERTAINTY

Uncertainty analysis for transmission/reflection measurements is challenging, and usually requires even more effort than the measurement itself. An inversion algorithm, which may include numerical root finding methods, calculates the permittivity and permeability. So the relationship between permittivity and permeability and the scattering parameters is not a simple functional relationship. The primary factors that contribute to uncertainty include the following:

- Inherent network analyzer accuracy limits

- Specimen position uncertainty

- Specimen dimension uncertainty

- Transmission line limitations
  (deviation from true plane wave, multipath noise)

There are several methods for adding up the total uncertainty from these contributions, ranging from rigorous Monte-Carlo calculations to analytic uncertainty estimates. Monte-Carlo methods have the advantage of more rigorously representing the functional dependencies by explicitly incorporating the governing equations that relate measured variables to the calculated parameters. However the Monte-Carlo method requires more computational resources since it must evaluate the governing equations repeatedly to obtain a statistically significant sampling. Analytical methods instead approximate the uncertainties with a function that reduces the need for multiple evaluations of the governing equations. However these analytical functions are approximate.

The Taylor series method is a commonly used analytical method for uncertainty calculation, but may include truncation errors depending on the linearity of the relationship between the measured variables and the calculated parameters. It does have the advantage of a very simple implementation relative to other methods so is reviewed in the following discussion. Other more advanced uncertainty propagation methods based on numerical methods are reviewed elsewhere [49].

An example of a Taylor series uncertainty estimate, also known as a root mean square RMS error analysis is as follows [50],

$$\delta\varepsilon = \sqrt{\begin{array}{c}\left(\frac{\partial\varepsilon}{\partial|S_{11}|}\delta|S_{11}|\right)^2 + \left(\frac{\partial\varepsilon}{\partial\Phi_{11}}\delta\Phi_{11}\right)^2 + \left(\frac{\partial\varepsilon}{\partial|S_{21}|}\delta|S_{21}|\right)^2 \\ + \left(\frac{\partial\varepsilon}{\partial\Phi_{21}}\delta\Phi_{21}\right)^2 + \left(\frac{\partial\varepsilon}{\partial t}\delta t\right)^2\end{array}} \qquad (5.1)$$

$$\delta\mu = \sqrt{\begin{array}{c}\left(\frac{\partial\mu}{\partial|S_{11}|}\delta|S_{11}|\right)^2 + \left(\frac{\partial\mu}{\partial\Phi_{11}}\delta\Phi_{11}\right)^2 + \left(\frac{\partial\mu}{\partial|S_{21}|}\delta|S_{21}|\right)^2 \\ + \left(\frac{\partial\mu}{\partial\Phi_{21}}\delta\Phi_{21}\right)^2 + \left(\frac{\partial\mu}{\partial t}\delta t\right)^2\end{array}} \qquad (5.2)$$

where the uncertainties in in ε and μ are calculated by a sum of squares evaluation. These equations apply to the NRW inversion or iterative S_{21} and S_{11} methods described above, where both transmission and reflection are used to determine the permittivity and permeability. The δ operator designates the standard uncertainty. The measured variables included in these sums are amplitude of the scattering coefficient, $|S_{ij}|$, the phase of the scattering coefficient, Φ_{ij}, and specimen thickness, t. The specific form of these equations will depend on which method is used to invert permittivity and/or permeability. In this error propagation method, the standard uncertainties are multiplied by a weighting factor, which is simply the derivative of the calculated quantity with respect to the measured variable.

The various weighting factors can be pre-computed by taking the relevant derivatives of the constitutive equations for the applicable inversion algorithm. An example of using this Taylor uncertainty estimate method is given for the case of waveguide and coaxial transmission lines in [42 and 45]. These analyses are easily adapted to the free-space measurement method.

The remainder of this section is devoted to outlining the dominant uncertainty sources in focused beam measurements. These sources can be used to build an uncertainty model that depends on the inversion algorithm being used, as well as the measurement hardware. Conversely, improving measurement accuracy is principally addressed by devising new strategies for reducing these uncertainty sources.

TRANSMISSION LINE ERROR

Because of mismatches at the various junctions in the system (i.e. horns, lenses, cables) undesired reflections will corrupt the measured scattering parameters. Time-domain measurements such as those shown in Figure 19 have shown that the two largest mismatches are usually the discontinuities at the network analyzer port and at the input to the antenna. Interference from these mismatch reflections must be removed with calibration and data processing to allow more accurate measurement of the specimen signals. Even with calibration, however, there still remains error in the measured signal.

This can be illustrated through a simplified transmission line model, as shown in Figure 28. This model describes the characteristics of a reflection (S_{11}) measurement, where for simplicity only one side of a focused beam system (i.e one horn and lens) are modeled. The focused beam model includes shunt impedances for the mismatches of the network analyzer port (Z_1), the antenna (Z_2), and the reflection from the specimen (Z_3). Z_0 is the intrinsic impedance of the transmission line, which is free space in this case, and is therefore approximately 377 ohms.

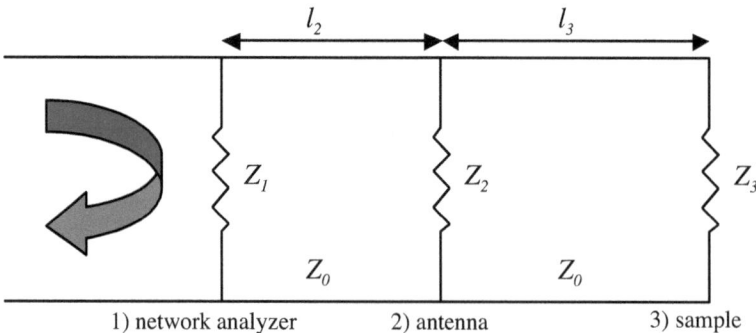

Figure 28   Transmission line model for estimating effects of horn and lens reflections.

This transmission line model can be applied to evaluate the effectiveness of the calibration. For illustration, we assume the response and isolation calibration method described in Chapter 2 by Equation (2.35). In reflection, this calibration technique uses two standards: an electrical short as the response ($S_{11} = 1$) and a matched load as the isolation ($S_{11} = 0$). The short is a metal plate with an area much greater than the incident beam, oriented normal to the incident beam; and the matched load is created by removing the metal plate and positioning absorber, such as pyramidal foam, approximately half a meter beyond the specimen position and tilting it slightly so that any residual specular reflection is redirected away from the receive antenna.

Referring to Figure 28, the reflection coefficient at the beginning of the transmission line (e.g. as measured by the network analyzer) due to the combined effects of the network analyzer port, antenna, and specimen can be computed by a cascaded matrix formulation. By assuming $Z_3 = 0$ for the short calibration standard and $Z_3 = Z_0$ for the load calibration standard, the reflection coefficients for those cases can be applied via Equation (2.35) to obtain,

$$S_{11}^{cal} = \frac{Z_3 - Z_0}{Z_3 + Z_0 + \zeta Z_3} \tag{5.3}$$

where,

$$\zeta = \frac{Z_0(e^{-ia} - e^{-ib}) - 2(Z_2 e^{-ia} + Z_1 e^{-ib})}{Z_0(\cos b - \cos a) + Z_1(e^{-ib} - e^{-ia}) - 2iZ_2 \sin a - 2\frac{Z_1 Z_2}{Z_0} e^{-ia}} \tag{5.4}$$

and where $a = \beta(l_3 + l_2)$, $b = \beta(l_3 - l_2)$, β is the propagation constant $(2\pi/\lambda)$, l_2 and l_3 are the network analyzer -- antenna and antenna -- specimen separations, and Z_0 is the line impedance.

The voltage-reflection coefficient of a load impedance, Z_3, in an ideal transmission line with line impedance Z_0, is given by,

$$S_{11}^{ideal} = \frac{Z_3 - Z_0}{Z_3 + Z_0} \tag{5.5}$$

Comparison of Equations (5.3) and (5.5) show that application of the response and isolation calibration does not completely eliminate the effects of the antenna and lens mismatches. There is a residual term, ζZ_3, which is proportional to the reflection properties of the specimen under test. Note that when minimizing the other mismatches, Z_1 and $Z_2 \to \infty$, ζ goes to zero and this term disappears. Thus minimizing the impedance mismatches at the network analyzer and antenna can improve measurement accuracy.

The amplitude of ζ can be estimated by assuming the antenna has a voltage standing wave ratio (VSWR) of approximately 2, so that the ratio of line impedance to the antenna impedance is, $Z_0/Z_2 \sim 0.5$. Assuming the contributions from other mismatches can be neglected compared to the antenna, then $|\zeta|$ can be as much as 0.25. With this value of $|\zeta|$, the ζZ_3 term can be significant and additional data processing must be used to reduce this systematic error and increase the accuracy of the calculated reflection coefficient. In other words, response and isolation calibration reduces the effects of the primary mismatch reflections, but does not eliminate the residual multi-path reflections that interact with the specimen. Because of this, time-domain gating is usually used to further refine the measured data, and a more sophisticated error model that directly simulates the time-domain gate effects is necessary.

FOCUSING ERROR

Determining the transmission and reflection coefficients from a material specimen assumes the incident energy is equivalent to an ideal, far-field, plane wave. In reality, the focused beam under-illuminates a specimen, providing near-constant phase and a tapered amplitude on the specimen. This Gaussian-like amplitude taper means the focused illumination is not a true plane wave but rather an approximation. To determine the effect of the finite beam width on transmission and reflection from a target, Petersson and Smith have compared the transmitted and reflected power to that from an infinite illumination case via plane wave spectrum analysis [51].

Figure 29 illustrates the impact of a tapered beam in system performance in terms of a plane wave spectrum representation. The incident field is expressed as a superposition of plane waves. This series of incident plane waves propagates through a planar specimen and the propagated plane waves are summed to determine transmitted and reflected power. Specifically, this scenario is analyzed in terms of the time-average power passing through input and output reference planes defined on either side of the specimen.

The plane waves are split into TE and TM (parallel and perpendicular) components, which are used to express the transmission and reflection coefficients (Fresnel coefficients). The following expressions for incident, transmitted, and reflected power, complete without any approximations, are then derived [52].

$$P_i = \frac{\pi}{4\eta_0} w_0^2 |E_0|^2 (k_0 w_0)^2 \int_{\tilde{k}_r=0}^{1} \left(1 - \frac{\tilde{k}_r^2}{2}\right) \frac{\tilde{k}_r}{\tilde{k}_z} e^{-\tilde{k}_r^2 (k_0 w_0)^2 / 2} d\tilde{k}_r \qquad (5.6)$$

$$P_t = \frac{\pi}{8\eta_0} w_0^2 |E_0|^2 (k_0 w_0)^2 \int_{\tilde{k}_r=0}^{1} \frac{\tilde{k}_r}{\tilde{k}_z} \left[\tilde{k}_z^2 |T_\perp(\tilde{k}_r)|^2 + |T_\parallel(\tilde{k}_r)|^2\right] e^{-\tilde{k}_r^2 (k_0 w_0)^2 / 2} d\tilde{k}_r \qquad (5.7)$$

$$P_r = \frac{\pi}{8\eta_0} w_0^2 |E_0|^2 (k_0 w_0)^2 \int_{\tilde{k}_r=0}^{1} \frac{\tilde{k}_r}{\tilde{k}_z} \left[\tilde{k}_z^2 |R_\perp(\tilde{k}_r)|^2 + |R_\parallel(\tilde{k}_r)|^2\right] e^{-\tilde{k}_r^2 (k_0 w_0)^2 / 2} d\tilde{k}_r \qquad (5.8)$$

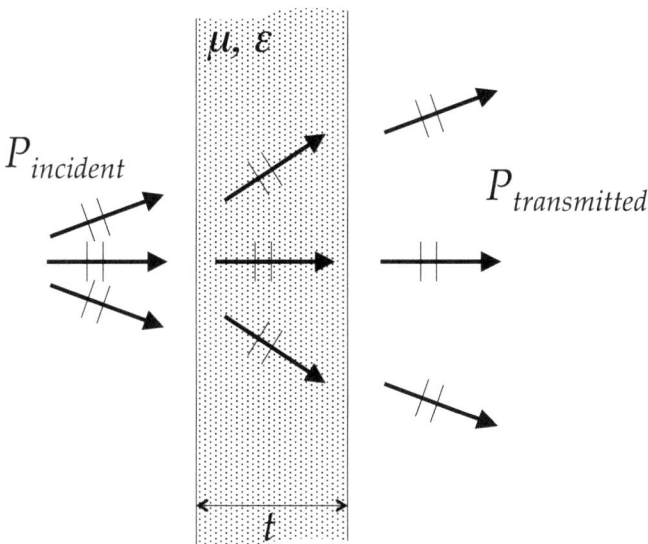

Figure 29   Sketch of problem geometry showing plane wave propagation through specimen.

where,

$$T_\perp(\tilde{k}_r) = \frac{4\tilde{k}_z\tilde{k}_{z\mu}e^{-i\tilde{k}_{zs}k_0 t}}{2\tilde{k}_z\tilde{k}_{z\mu} + \tilde{k}_z^2 + \tilde{k}_{z\mu}^2 + \left[2\tilde{k}_z\tilde{k}_{z\mu} - \tilde{k}_z^2 - \tilde{k}_{z\mu}^2\right]e^{-i2\tilde{k}_{zs}k_0 t}} \qquad (5.9)$$

$$R_\perp(\tilde{k}_r) = \frac{\tilde{k}_z^2 - \tilde{k}_{z\mu}^2 - \left[\tilde{k}_z^2 - \tilde{k}_{z\mu}^2\right]e^{-i\tilde{k}_{zs}k_0 t}}{2\tilde{k}_z\tilde{k}_{z\mu} + \tilde{k}_z^2 + \tilde{k}_{z\mu}^2 + \left[2\tilde{k}_z\tilde{k}_{z\mu} - \tilde{k}_z^2 - \tilde{k}_{z\mu}^2\right]e^{-i2\tilde{k}_{zs}k_0 t}} \qquad (5.10)$$

$$T_\parallel(\tilde{k}_r) = \frac{4\tilde{k}_z\tilde{k}_{z\varepsilon}e^{-i\tilde{k}_{zs}k_0 t}}{2\tilde{k}_z\tilde{k}_{z\varepsilon} + \tilde{k}_z^2 + \tilde{k}_{z\varepsilon}^2 + \left[2\tilde{k}_z\tilde{k}_{z\varepsilon} - \tilde{k}_z^2 - \tilde{k}_{z\varepsilon}^2\right]e^{-i2\tilde{k}_{zs}k_0 t}} \qquad (5.11)$$

$$R_\parallel(\tilde{k}_r) = \frac{\tilde{k}_z^2 - \tilde{k}_{z\varepsilon}^2 - \left[\tilde{k}_z^2 - \tilde{k}_{z\varepsilon}^2\right]e^{-i\tilde{k}_{zs}k_0 t}}{2\tilde{k}_z\tilde{k}_{z\varepsilon} + \tilde{k}_z^2 + \tilde{k}_{z\varepsilon}^2 + \left[2\tilde{k}_z\tilde{k}_{z\varepsilon} - \tilde{k}_z^2 - \tilde{k}_{z\varepsilon}^2\right]e^{-i2\tilde{k}_{zs}k_0 t}} \qquad (5.12)$$

and where,

$$\tilde{k}_z = \sqrt{1 - \tilde{k}_r^2}, \quad \tilde{k}_{zs} = \sqrt{\mu\varepsilon - \tilde{k}_r^2}, \quad \tilde{k}_{z\mu} = \frac{\tilde{k}_{zs}}{\mu}, \quad \tilde{k}_{z\varepsilon} = \frac{\tilde{k}_{zs}}{\varepsilon} \qquad (5.13)$$

and where t is the specimen thickness, $k_0 = 2\pi f/c$ is the free-space wavenumber, and μ and ε are the relative permeability and permittivity of the specimen. Note that the tilde symbol denotes when the k-vector is already normalized to k_0. The power reflection and transmission coefficients for the finite beam are then computed as,

$$R_{beam} = P_r/P_i \quad \text{and} \quad T_{beam} = P_t/P_i, \qquad (5.14)$$

which can then be integrated numerically. The relative error due to the finite beam is then determined by comparing these reflection and transmission coefficients to that from an ideal plane wave. The plane wave reflection and transmission power coefficients, R_{pw} and T_{pw} are computed from Equations (5.9) thru (5.12) when $\tilde{k}_r = 0$. The relative error in the scattering parameters due to the finite beam size is then,

$$\delta S_{11} = \frac{\sqrt{R_{beam}} - \sqrt{R_{pw}}}{\sqrt{R_{pw}}} \quad \text{and} \quad \delta S_{21} = \frac{\sqrt{T_{beam}} - \sqrt{T_{pw}}}{\sqrt{T_{pw}}} \qquad (5.15)$$

The relative error in the transmission coefficient is plotted for some representative cases in Figure 30 as a function of wavelength-normalized specimen thickness. Note that the wavelength normalization includes the effect of the permittivity and permeability on wavelength within the material. Three different curves are shown on each plot, each corresponding to a different dielectric permittivity of the specimen. These data show that for a beam radius of $k_0 w_0 = 8$, the systematic errors are a fraction of a percent, with the general trend for error levels to gradually increase with increasing electrical thickness of the specimen. Interestingly, error levels are lower when the specimen has a magnetic permeability that is greater than 1.

Figure 31 compares the relative transmission (top plot) and reflection (bottom plot) error as a function of electrical thickness for an $\varepsilon = 10$ dielectric slab with several different beam diameters ($k_0 w_0 = 6, 8,$ and 12). These data show that increasing the beam diameter reduces the systematic error from focusing. Note also that Figure 31 shows the relative error becomes exceedingly large in S_{11} at specimen thicknesses corresponding to integral multiples of $\lambda/2$. These errors are due to the inherent resonance caused by interference between the front and back surface of the specimen. Under this condition, S_{11} becomes increasingly small. For these examples, the plane-wave approximation results in systematic error less than 0.1 % for a $k_0 w_0 = 8$ focused beam system, similar to the one pictured in Figure 17.

SPECIMEN POSITION

Specimen positioning error primarily affects S_{11} measurements. Ideally the face of the specimen is either at exactly the same position or at a known displacement from the reference plane set by the response calibration. However if the calibration is not completely flat or the specimen is not completely flat, there can be uncertainty in this position. In this case, the transmission line path length varies by that uncertainty so that there will be a phase offset. Thus the uncertainty in phase angle is expressed by the uncertainty in the specimen position multiplied by the propagation constant,

$$\delta\theta = \gamma_0 \delta L, \quad \text{where} \quad \gamma_0 = \sqrt{k_c^2 - k_0^2} \qquad (5.16)$$

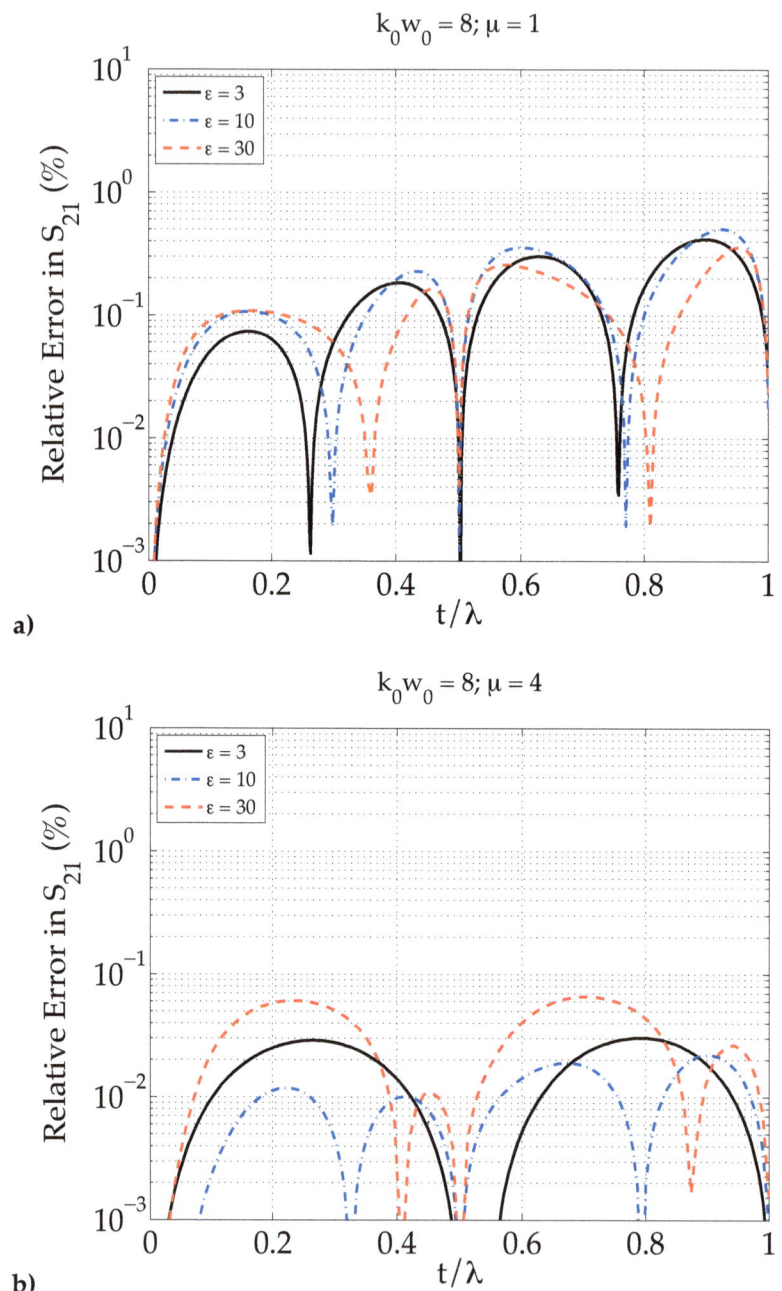

Figure 30 Relative error in plane-wave approximation for transmission when $k_0 w_0 = 8$: a) dielectric only slabs and b) magnetic slabs

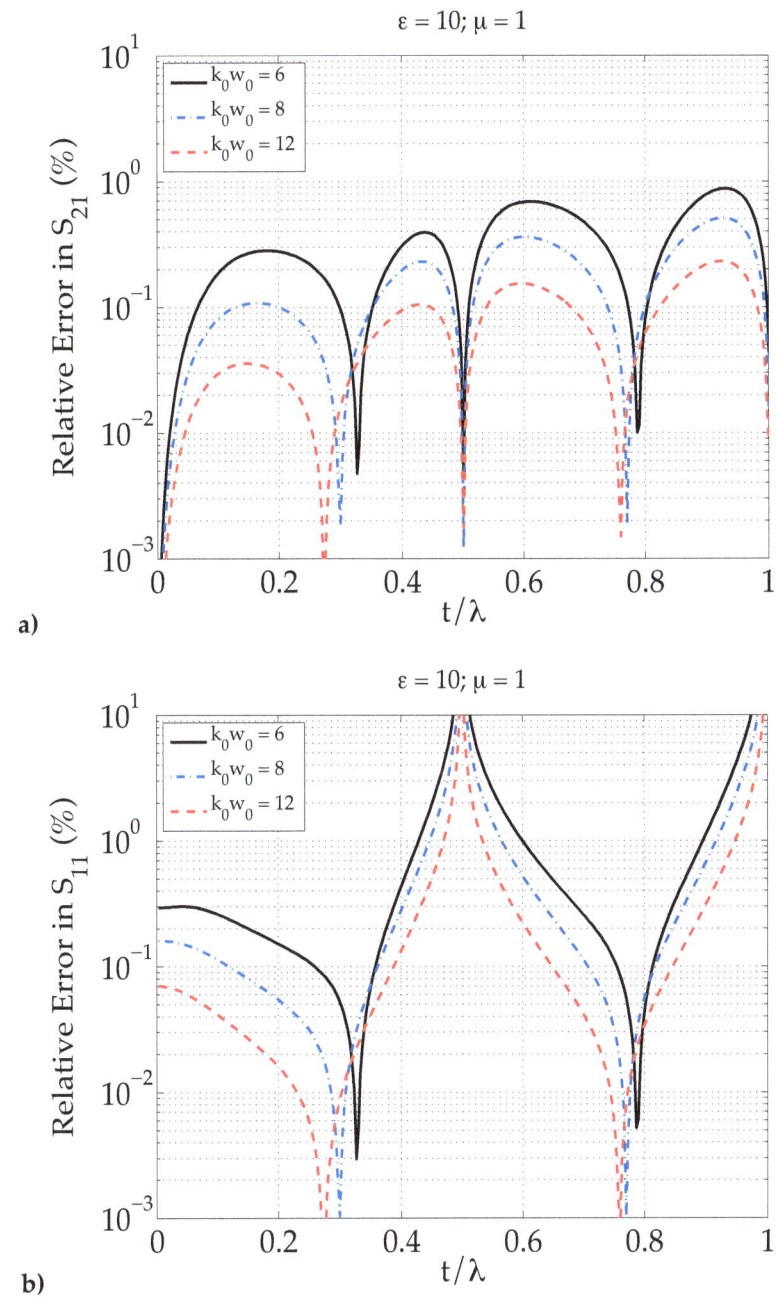

a)

b)

Figure 31   Relative error in plane-wave approximation with a dielectric slab for different beam diameters ($k_0 w_0$).

This source of uncertainty is not normally a concern. When measuring non-magnetic materials, the iterative S_{21} inversion algorithm is typically used, which is insensitive to the specimen position. When measuring materials where both permittivity and permeability are unknown, then the position independent, 4-parameter algorithm can be used, which is also insensitive to specimen position.

OTHER ERRORS: NETWORK ANALYZER & SPECIMEN

An overview of error sources in vector network analyzers is given by Rytting [53], and Wong provides a history of network analyzer calibration [54]. The error caused by limitations of the Network Analyzer is dependent on the specific equipment used, and manufacturers provide error estimates specific to their equipment. These errors will also depend on the calibration method and post-processing.

Another, often dominant source of uncertainty lies in the material specimen itself. A materials measurement laboratory may handle a wide variety of materials from various sources, including complex composite materials. The exact microstructure of these material samples may be unknown and there may be inhomogeneities that violate the assumptions of the inversion algorithms outlined above. In engineered composites, the material specimens are constructed from multiple constituents and there may be spatial variations within the specimen due to inconsistent distribution of these constituents. For example, in fiber-reinforced composites, there may be variations in the fiber weave or there may be trapped voids that lead to local variations of the dielectric properties. In mixtures, such as magnetic absorber materials or artificial dielectrics, which consist of magnetic or dielectric pigments within a polymer binder, there may be particle settling that leads to varying concentrations of pigment particles within the material.

An illustration of specimen inhomogeneity is shown in Figure 32, which shows a simple two-layer model of a material specimen, in which one side has a greater concentration of magnetic pigment than the other. As a result, one side has a significantly higher permittivity and permeability than the other side. When this two-layer material specimen is inverted as if it is homogeneous, then the resulting apparent permittivity and permeability of Figure 33 results. In this case, the material, as specified in Figure 32 was inverted using the four-parameter method. Of particular note is the imaginary part of the permittivity, which becomes negative at the higher frequencies of Figure 33. With the sign convention adopted in this book, imaginary permittivity and permeability should never be less than 0, so the apparent intrinsic properties of Figure 33 are not physically realizable and are therefore erroneous. Thus not only can the measurement fixture impose uncertainties on measured data, but so can the specimen itself, particularly under the assumption that it is homogeneous. Ideally, a materials measurement laboratory should have access to optical or electron microscopes so that specimen microstructures can be evaluated and correlated to the intrinsic microwave properties.

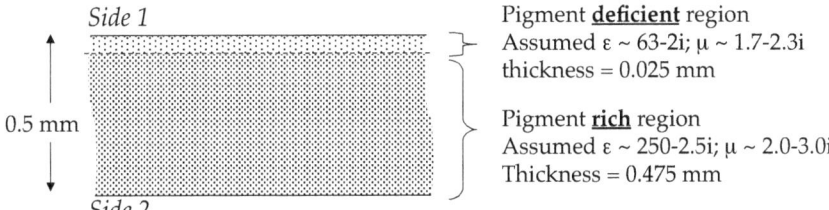

Figure 32   Diagram of representative inhomogeneity due to pigment particle settling in a composite mixture.

Another source of uncertainty, associated with the specimen, is the specimen thickness. Thickness errors may arise from the uncertainty in thickness estimation (i.e. micrometer measurements), or from surface roughness or thickness variations in the specimen itself. Specimen dimensions are typically measured with a caliper or micrometer and uncertainty in these instruments is typically less than 0.0005 inches (0.0127 mm). Since specimens may not be uniformly thick, it is important to measure the center of the specimen, and not just the edge; and this can be accomplished with a micrometer that has an extra deep yoke. When a specimen is centered in the focused beam system, most of the illuminating power is in the center of the specimen, so emphasis should be placed on the center thickness. Even when the center thickness is measured, there may be substantial variation in that thickness over the area of the beam diameter (> 15 square inches), which represents a source of uncertainty in thickness, t. Sometimes material specimens are elastomeric and therefore soft. In this case the measured thickness may depend on how much pressure is exerted by the micrometer or caliper during the thickness measurement.

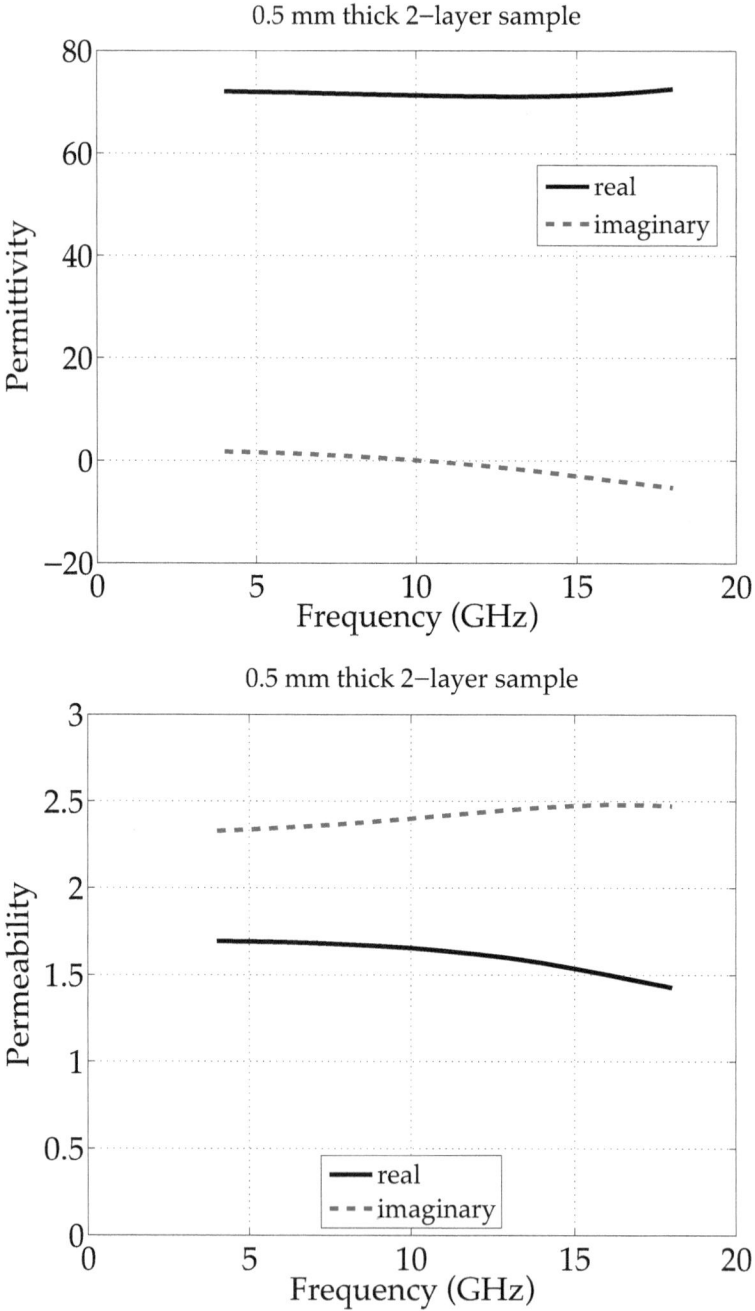

Figure 33   Apparent permittivity (top) and permeability (bottom) of
an inhomogeneous specimen assumed to be
homogeneous.

To illustrate the effect of thickness and network analyzer errors on the uncertainty in inverted properties, some representative cases are shown in Figure 34 and Figure 35. All these uncertainty data are shown as a function of the electrical specimen thickness. The thickness and network analyzer uncertainties were propagated using the root mean square method of Equations (5.1) and (5.2). In these plots, the thickness error was assumed to be ± 0.5 mil (± 0.0127 mm). The network analyzer error was assumed to be ± 0.2 degrees in phase, and ± 0.5% in amplitude plus an absolute amplitude error of 0.001. The absolute amplitude error is equivalent to a measurement dynamic range of 60 dB (e.g. a relative noise floor of -60 dB with respect to the maximum). For this illustration, no other uncertainty sources were included.

Figure 34 shows the estimated measurement uncertainty for a low-loss, low-dielectric material (e.g. an unfilled or "neat" polymer). The top plot is the uncertainty in the real part of the permittivity and the bottom plot is the uncertainty in the imaginary part of the permittivity for this material. Curves for two different inversion algorithms are shown: i) S_{21} iteration and ii) S_{21} & S_{11} iteration. The uncertainty calculated for the S_{21} & S_{11} iteration is the same as what would be calculated for the NRW inversion method as well. When the S_{21} iteration is used, the uncertainty generally decreases as specimen thickness increases. This follows since a thicker specimen has a greater effect on the transmitted phase, providing an improved signal to noise.

On the other hand, when the S_{21} & S_{11} iteration method is used, there is a very large uncertainty for specimen thicknesses corresponding to integral multiples of $\lambda/2$. While the uncertainty in the imaginary permittivity (bottom plot of Figure 34) is very large for this case, the imaginary permittivity itself is a small number to begin with ($\varepsilon'' = 0.01$). So the absolute error for this example is small relative to the real part of the permittivity. This kind of uncertainty in low-loss dielectric material measurements is similar to what is typically observed in other transmission-line methods, such as rectangular waveguide or coaxial airline fixtures.

A different example of the uncertainty estimate is shown for a lossy magnetic material in Figure 35. In these data, inversion via the S_{21} & S_{11} iteration is assumed since both permittivity and permeability must be evaluated. The error curves show the calculated uncertainties for both dielectric and magnetic properties as a function of electrical thickness. Because the values of the imaginary permeability and permittivity are larger than the low-loss dielectric, their relative uncertainties are correspondingly lower, even though their absolute uncertainties are similar.

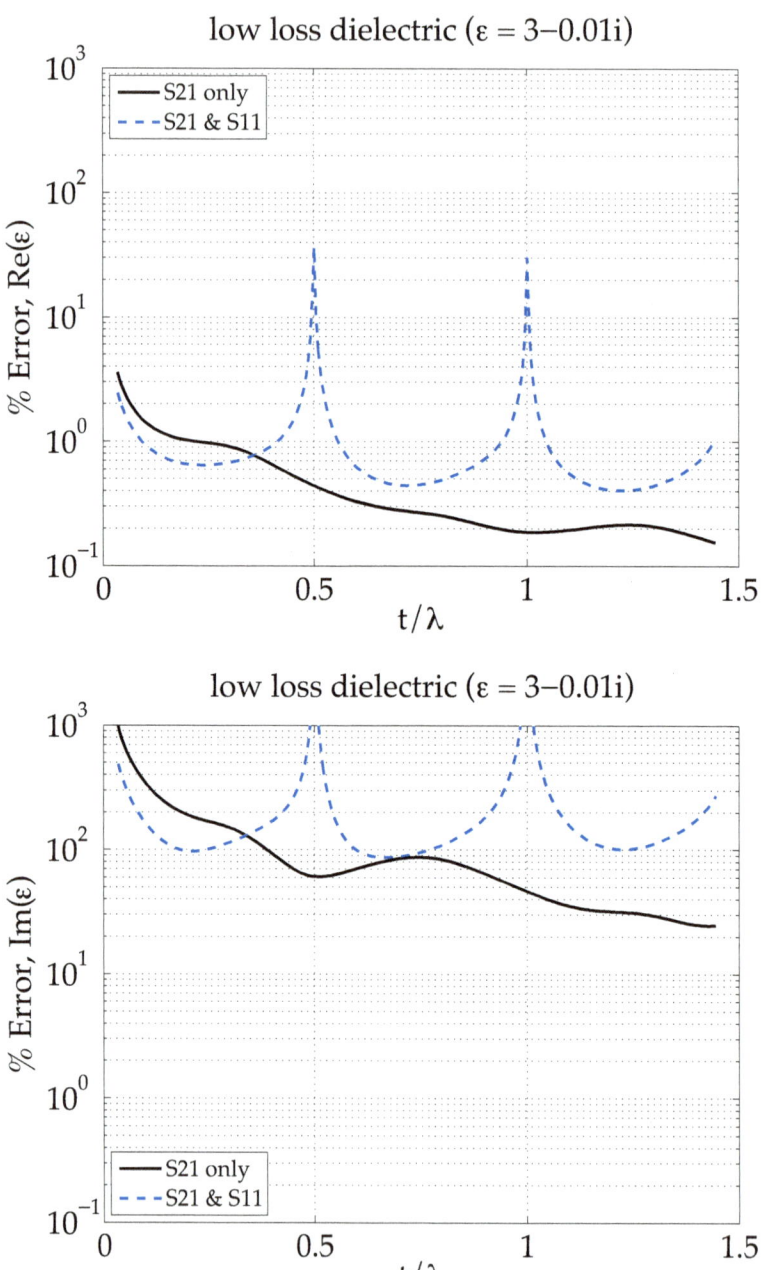

Figure 34   Representative measurement error for a low loss dielectric specimen due to thickness and network analyzer uncertainties.

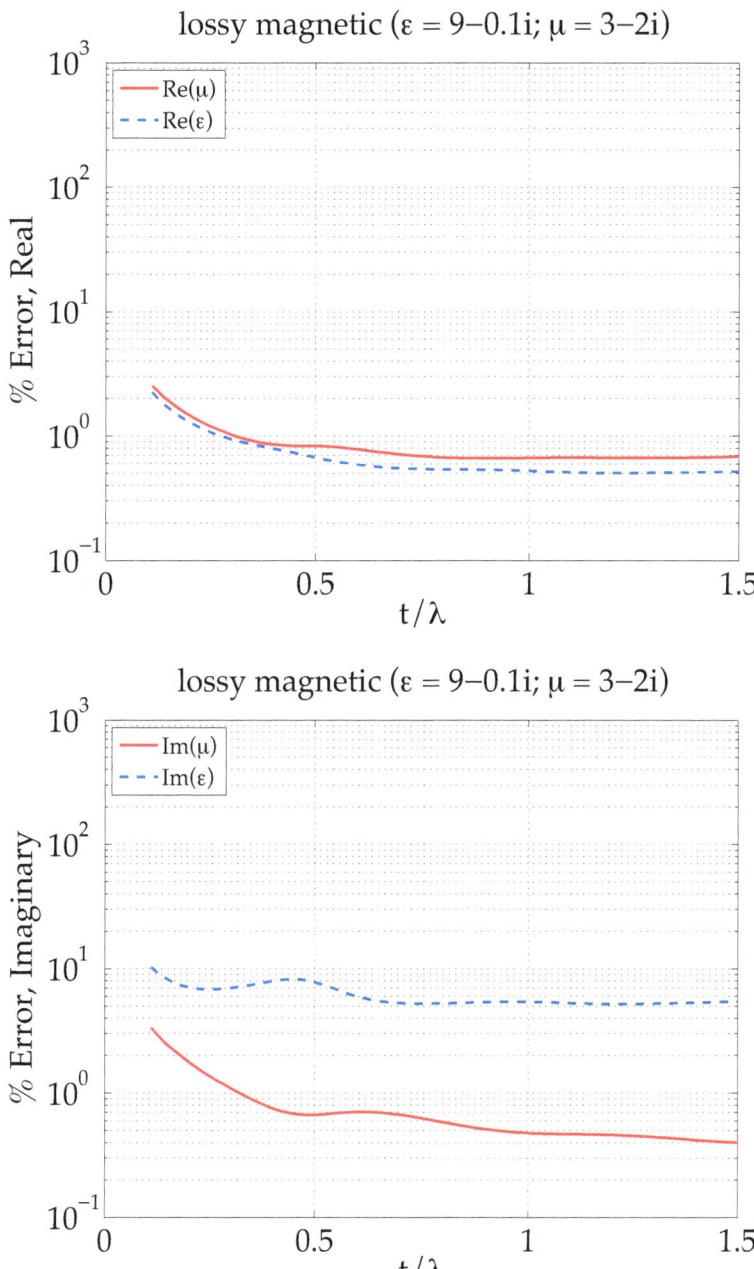

Figure 35   Representative measurement error for a lossy magnetic material due to thickness and network analyzer uncertainties.

6. MEASURING DIFFUSE SCATTER

BASIC CONCEPTS AND DEFINITIONS

So far, this book has discussed various aspects of measuring intrinsic properties of materials with free-space focused beam systems. Accurate determination of these intrinsic properties (e.g. permittivity and permeability) relies on measurement of specular scatter from a well-defined, homogeneous slab of material. The discussion in this chapter is about an alternative use of the focused beam system, where it is applied to characterize *diffuse* scatter that occurs in *in*homogeneous materials or structures. Specular scatter is defined by Snell's law, or more generally by the Fresnel equations [55], which describe the behavior of light at an interface in terms of complex reflection and refraction. The assumption of Snell's law is that the interface is smooth and homogeneous. However when electromagnetic energy encounters specimens with a rough surface, a portion of the incident energy may be scattered diffusely. At microwave frequencies, the difference between specular reflection and diffuse reflection is similar to the difference between a mirror and a painted wall at optical frequencies.

An example of bistatic microwave scatter, including diffuse and specular components is shown in Figure 36. The data shown in this figure were calculated with a two-dimensional finite difference time domain (FDTD) simulation of a 60 cm wide rough surface on an infinite ground plane. It was illuminated with a 10 GHz plane wave at 45 degrees incidence, and the roughness was randomly generated on a conductive ground plane with a peak-to-trough height of 3 millimeters (~ $\lambda/10$). The relative amplitude of the scatter in logarithmic units is shown in a polar plot. The radial scale is 10 dB per division, so the diffuse scatter varies from 10 to 30 dB below the specular lobe in this case. In typical engineered composites, the surface roughness is typically not this large. However in some materials, there can still be enough inhomogeneity to result in a small but nevertheless measureable diffuse scatter component.

Besides surface roughness, there are also other sources for diffuse scatter. For example a material surface may be smooth, but if it is a composite of two or more constituents, then the local dielectric or magnetic properties may vary spatially. Such an effect can be caused by variations in a composite weave or by variations in local volume loading of a pigment loaded polymer. Discontinuities such as an edge or a gap in the middle of a material specimen can also diffract energy causing diffuse scatter, and finally periodic arrays, whether in the form of a composite material like honeycomb, or as an engineered surface like a frequency selective surface or a Metamaterial [56, 57], can have geometric variations and imperfections. Periodic arrays can also exhibit coherent addition of the scatter from the array elements resulting in another type of scatter called grating lobes or Floquet scatter [57].

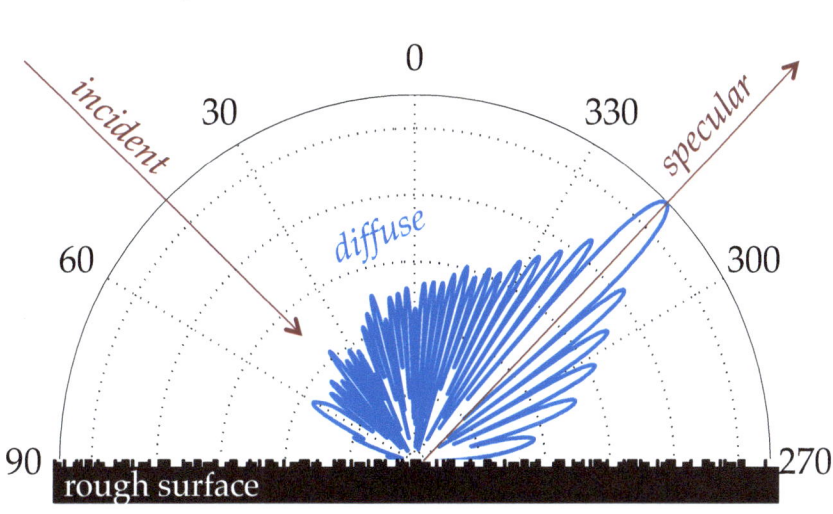

Figure 36   Calculated diffuse scatter from a 60 cm wide rough
surface at 10 GHz on an infinite ground plane.

The focused beam system provides a means with which to measure some of these non-specular forms of scatter and is particularly convenient for determining the monostatic backscatter portion of the diffuse "fuzz-ball". In this mode, a focused beam system can act as an alternative to a compact radar cross-section (RCS) measurement range [58]. Measuring microwave scattering, or RCS, from an object is usually a difficult and expensive prospect involving construction of complex models or test-bodies, and subsequent measurement in RCS range facilities. RCS ranges are typically large facilities that require significant levels of supporting manpower and equipment to obtain scatter data from a target under test. This is in contrast to a focused beam system, which is a smaller, laboratory-scale apparatus that can operate at a significantly lower-cost than an RCS range.

Before discussing the use of a focused beam to probe the scatter from an object, we must first quantify the concept of radar cross-section. The RCS of a complex object is a function of its size, shape, and material properties and is expressed in terms of an effective scattering cross-sectional area (e.g. square meters). Specifically, it is a measure of the reflective intensity of a target under radar illumination. When discussing the RCS at arbitrary directions, the term *bistatic* is often used to denote that the incidence and measurement angles are not necessarily the same. Bistatic RCS is what is plotted in the polar plot of Figure 36. However, when the direction of scatter under study is restricted to the direction back towards the illuminating radar, then the terms *monostatic* RCS or *backscatter* are used.

The RCS, σ, is defined as an effective cross-sectional area for an equivalent isotropic scatterer when illuminated by a plane wave. More specifically it is calculated by equating the captured power that is incident, to the power re-radiated by an isotropic scatterer. The captured incident power is equal to the effective cross-sectional area of the scatterer (i.e. the RCS) multiplied by the incident power per unit area: σW_i. The power re-radiated by an isotropic scatterer is the scattered power density at some distance, R, from the scatterer, multiplied by the surface area of a sphere with that same radius: $4\pi R^2 W_s$. Equating these two terms and rearranging to solve for RCS gives,

$$\sigma = 4\pi R^2 \frac{W_s}{W_i} \tag{6.1}$$

Because power density is proportional to the square of the electric or magnetic field, we can rewrite the ratio of the scattered and incident power densities in terms of fields. Furthermore, because of the dependence of this equation on R, we can standardize by taking the limit as R goes to infinity. Applying this to Equation 6.1, then gives the formal definition of radar cross section,

$$\sigma = \lim_{R \to \infty} 4\pi R^2 \left| \frac{E_s}{E_i} \right|^2 \qquad (6.2)$$

where E_i and E_s are the incident and scattered electric fields respectively. Note that this equation could alternately be written in terms of the scattered magnetic fields. Furthermore, the usual convention is to express this cross-section in a logarithmic scale: dB-square meters (dBsm), which is equal to $10 \log_{10} \sigma$, where σ is the radar cross-section in square meters.

Actual measurement of RCS requires an RCS measurement range facility to create a near-plane wave condition over some volume of space, called a "quiet zone". A target placed inside that quiet zone then scatters energy as if it were experiencing a plane wave. The energy scattered from that target is then used to determine the radar cross section. Since the measurement range has a finite size (i.e. R is not infinite), a calibration of a known target is used to calculate the desired RCS.

The return echo from an RCS measurement includes scattering from the edges and shape of the object, making it difficult to distinguish the individual contributions caused by a single discontinuity or by material roughness. Similar to an RCS range, a focused beam system also simulates a plane wave at the focal point of the beam. However in contrast to an RCS range, the focused beam only illuminates a small portion of a larger specimen or target. So a focused beam system can have the advantage of isolating the effects of individual discontinuities or material inhomogeneities from the scatter that comes from the target shape. In essence, it can focus on a particular portion of an object to distinguish between different scattering phenomenology and between individual scattering sources on a larger body.

The idea of a focused beam has also been applied to enable numerical electromagnetic simulations of rough surface scatter. Collin showed that Gaussian beam illumination could be used to approximate the scattering coefficient of an "infinite" rough surface as long as a planar phase front is maintained and the correlation lengths of the rough surface are smaller than the illuminating beam diameter [59]. While Collin emphasized the use of a tapered beam for numerical analysis, this chapter demonstrates the use of a Gaussian beam for experimental backscatter characterization.

When the scatter from a rough surface or inhomogeneous material is measured, it is quantified in terms of the RCS per unit area, or dBsm/sm;

which is designated by a subscript-zero, σ_0. This σ_0 is a commonly measured parameter in remote sensing applications [60], where radar backscatter of the planet's surface is mapped for environmental, agricultural, and other applications. σ_0 is sometimes called the "scattering coefficient". In addition, when characterizing the scatter from a linear discontinuity, such as a gap or an edge, the relevant scattering parameter is RCS per unit length [47]. This is also known as the echo width and is express in units of dBsm per meter, or more compactly as a two-dimensional RCS: dB-meters (dB-m).

SCATTERING COEFFICIENT MEASUREMENT

Unlike an RCS range, the focused beam illuminates only a portion of a target. Calibration is also different, in that a flat metal plate that extends outside the beam is used to establish a reflection reference. To derive the RCS of a scatter calibrated by this method, we can use a plane wave spectrum decomposition. Assuming the case of an x-oriented, linearly polarized field centered on the z-axis, the plane wave spectrum can be represented with a Fourier transform pair [61],

$$\frac{E_x(x,y,z)}{E_0} = \frac{1}{(2\pi)^2} \int_{-\infty}^{\infty} \int_{-\infty}^{\infty} F_x(k_x, k_y) e^{-i(k_x x + k_y y + k_z z)} dk_x dk_y \qquad (6.3)$$

$$F_x(k_x, k_y) = \int_{-\infty}^{\infty} \int_{-\infty}^{\infty} \frac{E_x(x,y,z=0)}{E_0} e^{i(k_x x + k_y y)} dx dy \qquad (6.4)$$

Additionally the scatter generated by a target within the illuminating beam has a spectrum defined by,

$$F_x^{scatter}(k_x, k_y) = \int_{-\infty}^{\infty} \int_{-\infty}^{\infty} F_0(k_x', k_y') R_s(k_x', k_y', k_x, k_y) dk_x' dk_y' \qquad (6.5)$$

where the spectrum of the incident beam is F_0, and R_s is the appropriate component of the plane wave scattering matrix [62], which relates the amplitude of the scattered plane wave in the direction (k_x, k_y) to a plane wave incident from the direction (k_x', k_y'). This scatter is then collected by a lens, which also imposes a spectral weight function, $F_x^{lens}(k_x, k_y)$. In a focused beam system, the scatter is measured by an equivalent lens and horn to that which generates the incident field. So reciprocity dictates that $F_x^{lens} = F_0$. The received signal is then obtained by multiplying the scattered spectrum with the spectral weight function of the lens,

$$\frac{V_r}{V_0} = \int_{-\infty}^{\infty} \int_{-\infty}^{\infty} F_x^{scatter}(k_x, k_y) F_0(k_x, k_y) dk_x dk_y \qquad (6.6)$$

The reflection calibration is referenced to a conductive metal plate that extends well beyond the extent of the beam and is centered at the beam waist, $z = 0$. Because it is an ideal reflector and effectively infinite in extent relative to the focused beam, the plane wave scattering matrix is described by $R_s(k_x', k_y', k_x, k_y) = \delta(k_x - k_x')\delta(k_y - k_y')$, where δ is the Dirac delta function. Applying this to Equation (6.5), and combining it with Equation (6.6), the received signal for the metal reference plate is then

$$\frac{V_{metal}}{V_0} = \int_{-\infty}^{\infty} \int_{-\infty}^{\infty} [F_0(k_x, k_y)]^2 dk_x dk_y \qquad (6.7)$$

Assuming that the focused beam closely approximates a plane wave and the phase taper is negligible, then $F_0(k_x, k_y)$ is approximately real. Furthermore, Parseval's theorem states that the integral of the square of a function is equal to the integral of the square of its Fourier transform. Using this, we can rewrite Equation (6.7),

$$\frac{V_{metal}}{V_0} = (2\pi)^2 \int_{-\infty}^{\infty} \int_{-\infty}^{\infty} \left| \frac{E_x(x, y, z = 0)}{E_0} \right|^2 dx dy = (2\pi)^2 A_{eff} \qquad (6.8)$$

where we note that the integral of the power is equivalent to the effective illumination area of the focused beam. When the beam follows a Gaussian amplitude taper, this is equivalent to the area in which the power density is within $1/e$ of the peak.

$$A_{eff} = \int_{-\infty}^{\infty} \int_{-\infty}^{\infty} e^{-x^2/w_x^2} e^{-y^2/w_y^2} dx dy = \pi w_x w_y \qquad (6.9)$$

This corresponds to the area of an ellipse defined by the horizontal and vertical beam waists.

RCS is defined in terms of an equivalent isotropic scatterer. With a similar plane wave analysis, we can also compute the scatter from a small, isotropic scatterer within the focused beam. A small target located at a position (x_0, y_0, z_0) within the beam will have a scattering matrix that can be approximated by,

$$R_s\left(k'_x, k'_y, k_x, k_y\right) \approx R_s(0)e^{-i\left(k'_x x_0 + k'_y y_0 + k'_z z_0\right)}e^{-i\left(k_x x_0 + k_y y_0 + k_z z_0\right)} \qquad (6.10)$$

Unlike the scattering matrix for the infinite plate reflector, the plane waves emanating from a small scatter show an exponential decay with distance from that scatterer. As in the case of the infinite metal plate reflector shown above, we can combine the scattering matrix of the small isotropic scatterer with Equation (6.5) and (6.6) to obtain

$$\frac{V_s}{V_0} = R_s(0)\left[\int_{-\infty}^{\infty}\int_{-\infty}^{\infty} F_0\left(k_x, k_y\right)e^{-i\left(k_x x_0 + k_y y_0 + k_z z_0\right)}dk_x dk_y\right]^2 \qquad (6.11)$$

From Equation (6.3), Equation (6.11) can be re-written in terms of the incident electric field,

$$\frac{V_s}{V_0} = R_s(0)\left[(2\pi)^2\frac{E_i(x_0, y_0, z_0)}{E_0}\right]^2 \qquad (6.12)$$

From Equations (6.12) and (6.8), and assuming that the scatterer is centered at the origin, we can now calculate the scattered signal relative to the metal plate,

$$\frac{V_s}{V_{metal}} = \frac{(2\pi)^2 R_s(0)}{A_{eff}} \qquad (6.13)$$

We can also write general expressions for the incident and scattered fields. Since we have placed the scatterer at the origin, $x = y = z = 0$, the following incident field is derived from Equation (6.3),

$$\frac{E_i}{E_0} = \frac{1}{(2\pi)^2}\int_{-\infty}^{\infty}\int_{-\infty}^{\infty} F_0\left(k_x, k_y\right)dk_x dk_y \qquad (6.14)$$

The scattered field can be written in terms of the spectral representations of three-dimensional far-field radiated fields [61].

$$\frac{E_s}{E_0} = \frac{ik_0}{2\pi r}F_x^{scatter}\left(k_x = 0, k_y = 0\right)e^{-ik_0 r} \qquad (6.15)$$

where Equation (6.5) tells us that

$$F_x^{scatter}(0,0) = \int_{-\infty}^{\infty} \int_{-\infty}^{\infty} F_0(k_x, k_y) R_s(k_x, k_y, 0, 0) dk_x dk_y \qquad (6.16)$$

These equations can then be applied to Equation (6.3) to get the radar cross section (RCS),

$$\sigma = 4\pi r^2 \left|\frac{E_s}{E_i}\right|^2 = 4\pi \frac{\frac{k_0^2}{(2\pi)^2} |F_x^{scatter}(0,0)|^2}{\frac{1}{(2\pi)^4} |\int_{-\infty}^{\infty} \int_{-\infty}^{\infty} F_0(k_x, k_y) dk_x dk_y|^2} \qquad (6.17)$$

If we approximate the incident field as a single plane wave, the spectral weight function is given by, $F_0(k_x, k_y) = \delta(k_x)\delta(k_y)$ and $F_x^{scatter}(0,0) = R_s(0)$. Then the RCS simplifies to $\sigma = 2(2\pi)^3 k_0^2 |R_s(0)|^2$. Combining this with Equation (6.13) then results in,

$$\sigma = 4\pi \frac{A_{eff}^2}{\lambda^2} \left|\frac{V_s}{V_{metal}}\right|^2 \qquad (6.18)$$

Thus this measured and calibrated voltage-reflection coefficient (e.g. $S_{11} = V_s/V_{metal}$) depends not only on the scattering properties of the material under test, but also on the illumination area of the incident beam.

The relationship between the voltage-reflection coefficient and RCS can also be derived more intuitively by assuming the reflection from the response calibration (metal plate) is equivalent to the RCS of a flat metal plate. In this assumption, the metal plate has an effective surface area equal to the illumination area of the beam (in the high frequency limit), $|V_{metal}|^2 \propto 4\pi(A_{eff}^2/\lambda^2)$. Then the calibrated scatter from the sample, $|V_{cal}|^2$, is equal to the ratio of the sample RCS to the metal plate RCS. With algebraic rearrangement, the sample RCS is expressed by,

$$\sigma = 4\pi \frac{A_{eff}^2}{\lambda^2} |V_{cal}|^2 \qquad (6.19)$$

which is equivalent to Equation (6.18). In traditional compact range measurements, calibration is accomplished by comparing the scatter of the target under test to that of a known standard shape, such as metal sphere or squat cylinder. The advantage of Equation (6.19) is that it is referenced to a metal plate, which is the usual calibration standard for focused beam reflection measurements. It doesn't require any additional calibration targets, and cylinders or spheres can be used instead as verification standards.

For an inhomogeneous surface, the quantity of interest is not RCS, but scattering coefficient. Instead of a target illuminated by a larger plane wave in an RCS range, a tapered plane wave is used to illuminate a larger specimen in the focused beam system. Because the target in this case is effectively infinite in size relative to the beam, the scattering coefficient (σ_o) is defined as RCS per unit physical area, where the illuminated (physical) area is the projected area divided by $\cos\theta$,

$$\sigma_0 = \frac{\sigma}{A_{eff}/\cos\theta} = 4\pi\cos\theta\,\frac{A_{eff}}{\lambda^2}\,|V_{cal}|^2 \tag{6.20}$$

where θ is the angle between the surface normal and the direction of the illuminating beam.

While the equations above were derived relative to a flat metal plate at normal incidence, this simple response calibration still leaves considerable noise in the signal due to other undesired reflections within the system. This is akin to the effects of background clutter in a traditional RCS measurement. Typical RCS calibration methods also include a background subtraction step to minimize the effects of other scatterers within the measurement range. In focused beam measurements, this is equivalent to the response and isolation calibration described previously. In particular, an isolation measurement, made with no metal plate or specimen present, is vector subtracted from both the response signal and from the specimen signal.

$$V_{cal} = \frac{S_{11}^{specimen} - S_{11}^{background}}{S_{11}^{metal\,plate} - S_{11}^{background}} \tag{6.21}$$

where S_{11} designates the monostatic reflection scattering parameter measured for the specimen, background, or metal plate. As in the case of most RCS measurement ranges as well as in the focused beam material measurements described previously in this book, time-domain processing (windowing or gating) is also used to further minimize errors from background scatter.

EXAMPLES OF SCATTERING COEFFICIENT MEASUREMENT

When a focused beam system is first configured for scattering coefficient measurement, a validation measurement of an ideal specimen should be performed to establish the minimum measurable backscatter levels. This validation also ensures that the system is operating within normal specifications. An ideal specimen in this case is a flat metal plate that should have negligible diffuse backscatter. An example measurement of this metal plate is shown in Figure 37. The data shown in this figure were for an aluminum plate that was 61 cm on a side, which corresponds to approximately 20 wavelengths at 10 GHz. The plate was rotated azimuthally so that monostatic reflection data were taken as a function of both frequency and angle. Thus the plate was much larger than the illuminating beam, minimizing the effects from edge diffraction, except at the most extreme angles. In addition, an absorber fence constructed from pyramidal foam absorber and placed around the measurement system minimized secondary reflections from objects in the room.

The measured backscatter at three different frequencies is shown at the top of this figure as a function of monostatic angle, while full angular data are shown in the lower part of the figure as a color image. Near zero degrees in azimuth, a lobe is evident due to the specular reflection of the beam at normal incidence. The amplitude and width of this specular lobe is a function of the illumination beam width, and meaningful non-specular (i.e. diffuse) backscatter is only measurable outside this lobe (beyond 20 degrees). As the beamwidth increases, the plane wave distribution narrows and this lobe also narrows. The non-specular scatter levels are as low as −65 dBsm/sm, indicating the sensitivity limit of the measurement apparatus. The residual signal measured below this level is due to internal network analyzer noise and the limits of the calibration methodology for eliminating multipath reflections. At angles near grazing the edges of the finite size plate begin to interact with the main part of the beam, causing increased scatter.

With a reasonable noise floor thus established, the backscatter of various specimens of interest can then be characterized. Figure 38 is an example of such a measurement. The specimen was 24 x 48 inches in dimension (61 cm by 122 cm) and consisted of a random arrangement of metal spheres arranged in a two-dimensional plane. The spheres consisted of metal BBs and were 0.173 inches (4.4 mm) in diameter. A total of approximately 6000 spheres were used to make the random surface. To contain the spheres, they were sandwiched between two pieces of low-dielectric foam and anchored with adhesive. Spray adhesive was used on each of the Styrofoam substrates to ensure the spheres were anchored. The random pattern was generated by sprinkling the spheres onto the adhesive covered foam from a height of approximately 60 inches. Then the second adhesive covered foam substrate was pressed on top of the spheres and the edges were taped to hold the entire assembly together. A photograph of the arrangement of the spheres before the top foam layer was glued in place is shown in Figure 39.

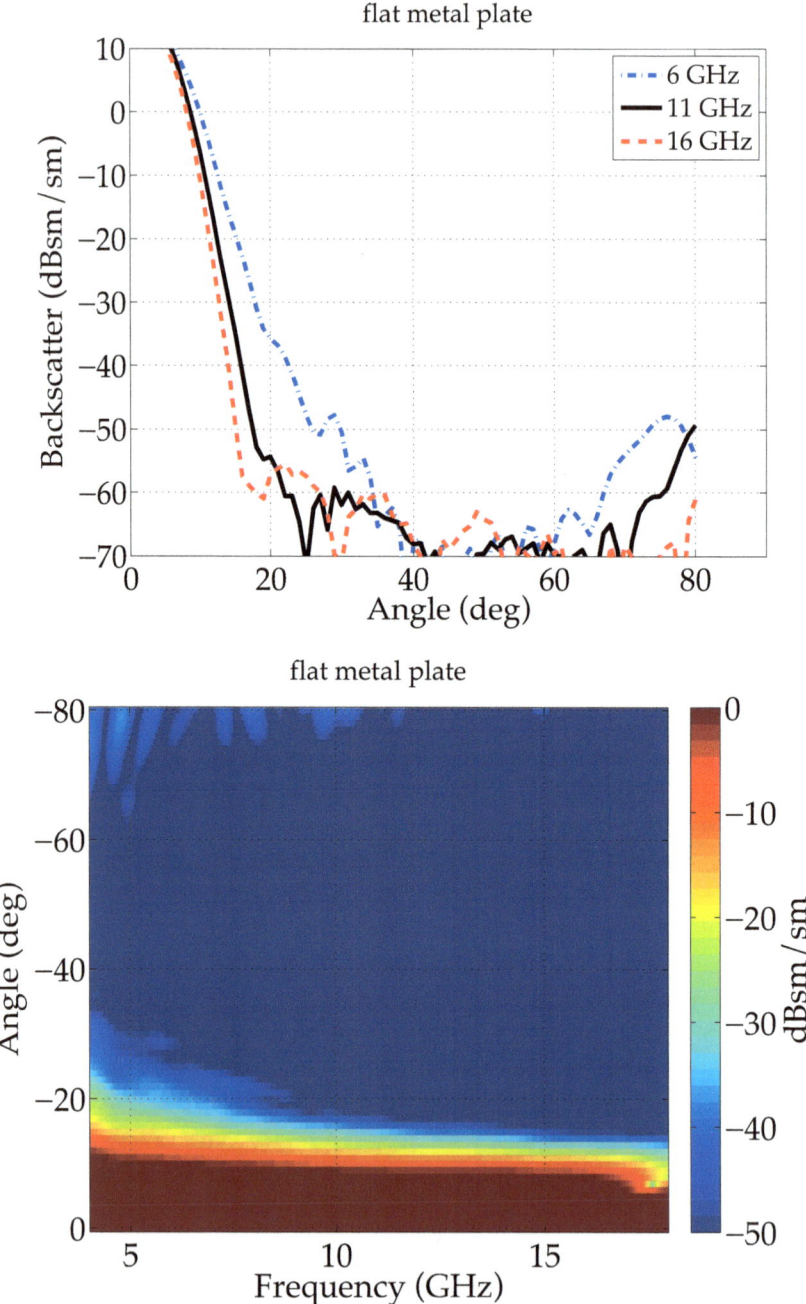

Figure 37   Diffuse scatter measurement of a flat metal plate
             displayed two different ways: angle cuts at 6, 11, and 16
             GHz (top), and full 2-D plot versus frequency and angle.

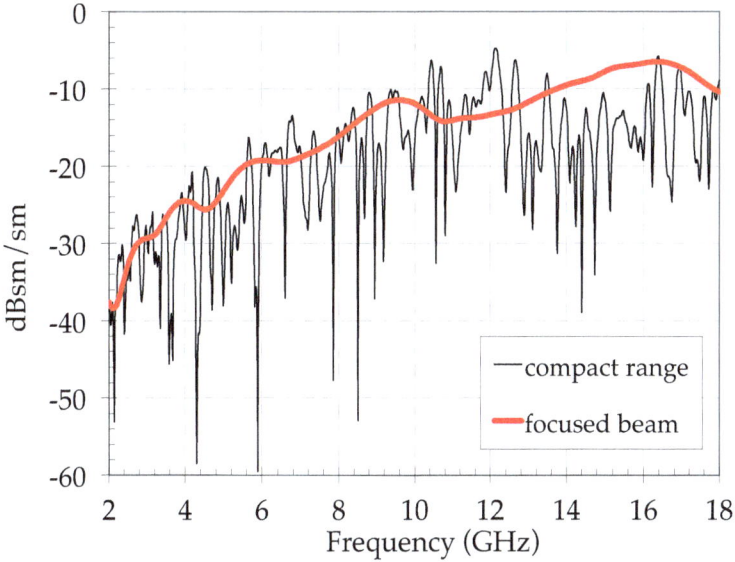

Figure 38 Comparison of the RCS per unit area of an inhomogeneous 24 x 48 inch (61 x 122 cm) specimen of random metal spheres.

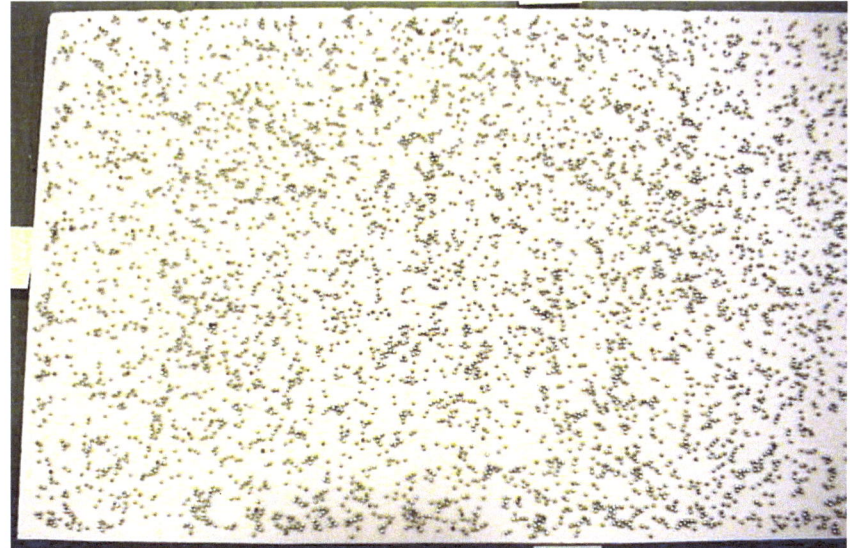

Figure 39 Photograph of a portion of BB-panel, in which a random arrangement of metal spheres was glued onto a foam substrate. The panel was 24 inches tall by 48 inches wide (61 x 122 cm).

This model rough surface was measured in both the compact RCS range and in the focused beam system with the rough scattering methodology developed above. Figure 38 shows a comparison of these data at a monostatic incidence angle of 45 degrees. Because of the limited beam width, the focused beam system was used to measure at different locations on the panel and the data shown are an average of measurements made at five different locations across the panel. While the compact range data follows the same trend as the focused beam data, it contains more fine structure. This effect is presumably due to interference effects from edge illumination in the compact range, which does not occur as strongly with the focused beam methodology. Other than the edge-diffraction effects, the data in Figure 38 show general agreement between the two measurement methods.

The focused beam method can also be applied to periodic materials such as a frequency selective surface (FSS). These FSS materials typically consist of a periodic pattern of conductive metal etched on a low-loss dielectric substrate. Figure 40 shows the Ka-band (26.5 to 40 GHz) backscatter measured from a rectangular array of Jerusalem-cross shaped slots [57] etched in a thin sheet of copper and supported on a dielectric substrate. The polarization of the incident beam was with the magnetic field transverse to the plane defined by the surface normal and the incident beam direction – TM. The periodicity of the Jerusalem cross patterning was 6.1 mm (0.24 inches). The main feature evident in Figure 40 is the grating lobe that extends from approximately 69 degrees at 26 GHz to 38 degrees at 40 GHz. The angular direction of this grating lobe is accurately described by diffraction theory and a simple equation can be derived,

$$2\pi(\sin\theta_r + \sin\theta_i) = \pm 2m\pi\frac{\lambda}{d} \qquad (6.22)$$

where $\theta_r = \theta_i$ for monostatic backscatter, $m = 1$, and d is the lattice constant of 6.1 mm. This equation applies to rectangular arrays and shows that grating lobes first appear when periodicity of the array corresponds to one half of the incident wavelength. The other strong feature in these data is the specular lobe that occurs at an angle of 0 degrees at all frequencies. As described previously, the width of this lobe is determined by the plane wave spectrum width, which is inversely related to the beam waist at the focus.

Other sources of diffuse scatter also exist in these data, albeit at lower amplitudes. These other diffuse scatter effects include variations in the Jerusalem cross lattice as well as inhomogeneity in the underlying substrate, which is a glass fiber reinforced dielectric composite.

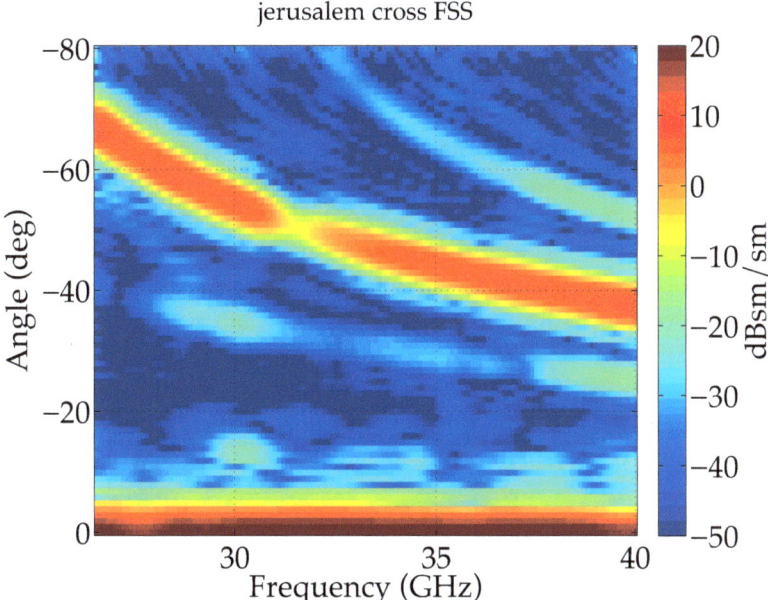

Figure 40 Diffuse scatter measurement of a Jerusalem-cross slot FSS versus frequency and angle. The data show a strong grating lobe that spans from 69 to 38 degrees in the measured frequency band.

ECHO WIDTH MEASUREMENT

The focused beam backscatter methodology can also be applied to linear discontinuities such as edges, cylinders, joints, or slots. In this case, the scattering figure of merit is echo width, measured in dB-meters. In terms of incident and scattered fields, echo width or two-dimensional RCS is defined by,

$$\sigma_{2D} = \lim_{\rho \to \infty} 2\pi\rho \left| \frac{E_{scattered}}{E_{incident}} \right|^2 \tag{6.23}$$

where ρ is the distance from the scattering discontinuity.

For a two-dimensional scatterer oriented along the y-direction and positioned at $x, z = 0$, the illuminating beam propagates in the z-direction with the beam waist centered at x, z, and $y = 0$. Since the receiver orientation is also positioned at $y = 0$ and the target is rotated about the y-axis, there is symmetry across the x-z plane. The contribution to the y-component of the received field from the $y > 0$ side of the target is cancelled by the $y < 0$ contribution. Thus the measured backscatter from these linear targets can be described as two-dimensional. This dictates that the measurement can only be made when the axis of the edge and the propagation vector are orthogonal (i.e. 0 degrees elevation).

As before, the sample scatter is referenced to a normal incidence metal plate. However, the echo width is calculated by assuming the normal incidence metal plate is equivalent to a two-dimensional strip with its width corresponding to the $1/e$ width of the beam. This $1/e$ width is a direct result of the assumption that the incident beam follows a Gaussian amplitude profile. Along similar lines to the derivation provided previously for the RCS per unit area, the echo width of a metal strip at normal incidence is given by, $2\pi(W_{eff}^2/\lambda)$, where W_{eff} is the illuminating beam width. Then the echo width of the sample is given by,

$$\sigma_{2D} = \frac{2\pi W_{eff}^2}{\lambda} |V_{cal}|^2 \qquad (6.24)$$

The disadvantage of this method is that it requires accurate characterization of the beam profile. An alternative method may also be used to calibrate, which instead uses known calibration standards, much like traditional RCS calibration in a compact range. In particular, long cylindrical rods can be used as standards, by insuring that the rod extends well beyond the illumination area. In this way, the rod appears to be effectively infinite in length, and an analytical expression for the echo-width of an infinite conductive (metal) cylinder can be applied. In logarithmic units, the calibrated echo width is computed from the return loss by,

$$\sigma_{2D}^{target} = V_{cal}^{target} + \sigma_{2D}^{cylinder} - V_{cal}^{cylinder} \qquad (6.25)$$

where V_{cal} is the return loss data, calibrated by the method describe previously in Equation 6.21. The value of $\sigma_{2D}^{cylinder}$ depends on the diameter of the metal cylinder and can be calculated analytically [47].

EXAMPLES OF ECHO WIDTH MEASUREMENT

An example application of this calibration equation is shown in Figure 41, which schematically shows the measurement geometry of non-specular backscatter caused by diffraction from a simple wedge. The scatter from an infinite edge can be derived analytically [63]. For the TE polarization (E vector is parallel to edge axis), and ignoring the specularly reflected components, the diffraction component of the electric field is given by the following equation,

$$E_z^{TE} = \frac{\pi}{\gamma} \sin\left(\frac{\pi^2}{\gamma}\right) \left[\frac{1}{\cos\left(\frac{\pi^2}{\gamma}\right) - 1} - \frac{1}{\cos\left(\frac{\pi^2}{\gamma}\right) - \cos\left(2\phi\frac{\pi}{\gamma}\right)} \right] \frac{e^{-i\rho k_0} e^{i\pi/4}}{\sqrt{2\pi\rho k_0}} \quad (6.26)$$

where $k_0 = 2\pi/\lambda_0$ is the wave number, γ is the exterior angle of the wedge, and ϕ is the incident angle, as shown in Figure 41.

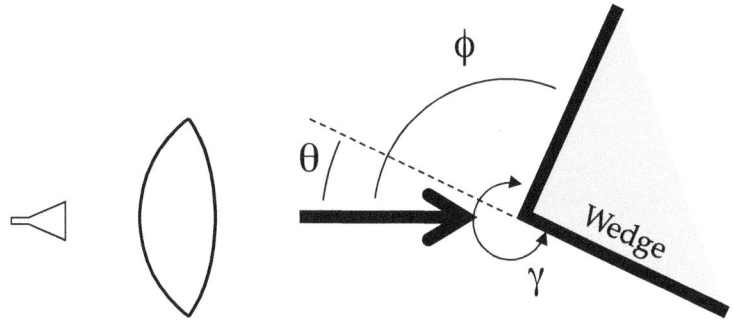

Figure 41   Schematic of measurement geometry for echo-width of a wedge.

Similarly for the TM polarization (E vector is perpendicular to edge axis), the magnetic field due to diffraction is given by,

$$H_z^{TM} = \frac{\pi}{\gamma} \sin\left(\frac{\pi^2}{\gamma}\right) \left[\frac{1}{\cos\left(\frac{\pi^2}{\gamma}\right) - 1} + \frac{1}{\cos\left(\frac{\pi^2}{\gamma}\right) - \cos\left(2\phi\frac{\pi}{\gamma}\right)} \right] \frac{e^{-i\rho k_0} e^{i\pi/4}}{\sqrt{2\pi\rho k_0}} \quad (6.27)$$

Using the definition for two-dimensional RCS, the TE and TM echo widths are then,

$$\sigma_{2D}^{TE} = \frac{1}{k_0} \left\{ \frac{\pi}{\gamma} \sin\left(\frac{\pi^2}{\gamma}\right) \left[\frac{1}{\cos\left(\frac{\pi^2}{\gamma}\right) - 1} - \frac{1}{\cos\left(\frac{\pi^2}{\gamma}\right) - \cos\left(2\phi\frac{\pi}{\gamma}\right)} \right] \right\}^2 \qquad (6.28)$$

and

$$\sigma_{2D}^{TM} = \frac{1}{k_0} \left\{ \frac{\pi}{\gamma} \sin\left(\frac{\pi^2}{\gamma}\right) \left[\frac{1}{\cos\left(\frac{\pi^2}{\gamma}\right) - 1} + \frac{1}{\cos\left(\frac{\pi^2}{\gamma}\right) - \cos\left(2\phi\frac{\pi}{\gamma}\right)} \right] \right\}^2 \qquad (6.29)$$

The diffraction echo width predicted by this classical diffraction theory is compared to a focused beam-measured and calibrated echo width (at 10 GHz) for the ninety-degree wedge in Figure 42. In particular, a ninety-degree metal wedge, with 45 cm wide sides was centered at the focus of the beam. The wedge was mounted on a turntable and backscatter was measured as a function of frequency and azimuth angle. The data at 10 GHz are plotted as a function of angle, θ, where $\theta = \phi - 90$ as defined in Figure 41. These data show quantitative agreement for most of the angles measured. Poor agreement exists near $\theta = 0$, where the measured scatter includes a specular component due to the normal incidence of one side of the wedge, and where the classical diffraction theory no longer applies. At angles near this 0-degree specular angle, the effects of finite beamwidth are also evident.

CROSS-POLARIZED SCATTER

Anisotropic materials and inhomogeneities can contribute a cross-polarization component to a scattered signal, particularly when the inhomogeneity is asymmetrical. Measuring this component with a free space system requires correcting for the cross-polarized contributions of all error sources, including the transmit and receive coupling errors of the feed antenna. It is possible to utilize a response and isolation calibration methodology for calibrating cross-polarized measurements based on a wire grid polarizer. In particular, the analytical expressions for reflection of parallel and perpendicularly polarized energy from a wire grid polarizer are as follows [64],

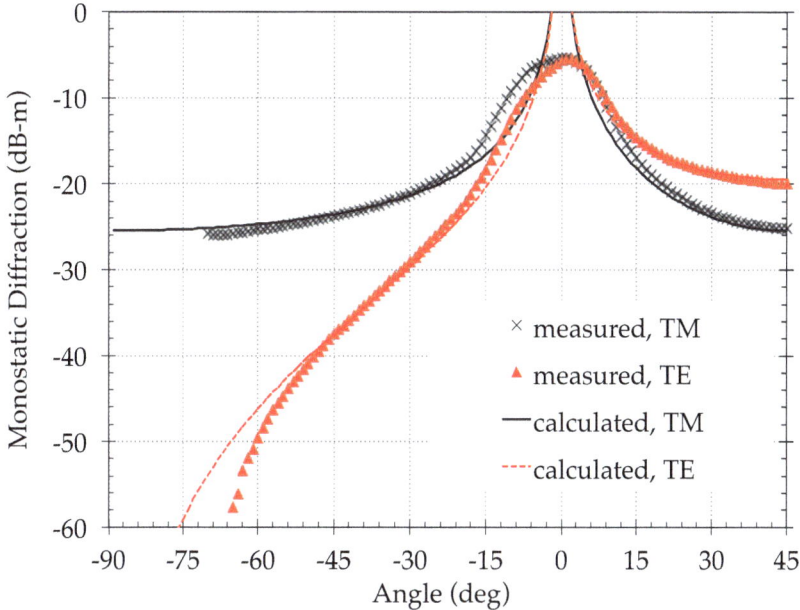

Figure 42   **Comparison of measured and calculated diffraction from a 90-degree wedge at 10 GHz.**

$$R_{power}^{\|} = \frac{1}{1 - \left(\frac{2d}{\lambda} \ln\left(\frac{d}{\pi a}\right)\right)^2}$$ (6.30)

$$R_{power}^{\perp} = \frac{\left(\frac{\pi^2 a^2}{2\lambda d}\right)^2}{1 - \left(\frac{\pi^2 a^2}{2\lambda d}\right)^2}$$ (6.31)

where the polarizer is made up of a periodic array of wires of radius, a, that are spaced with a periodicity of d. When the polarizer is rotated within the measurement plane to an angle of 45 degrees, the reflection of the polarizer in both polarizations is approximately -6 dB. The exact reflection coefficient is calculated from the above equations and used as a relative standard for normalizing the raw reflection loss. In a typical dual-pol horn antenna, the cross-pol isolation is typically only 20 to 30 dB below the co-pol signal; and improved isolation may be desired for some measurements.

A more accurate, full-polarimetric calibration methodology, based on a dihedral calibration standard, can be alternatively used. This cross-pol measurement methodology depends on a dual polarized antenna to enable simultaneous detection of two orthogonal polarizations. Measurements are performed with a vector network analyzer with Port 1 connected to the horizontal polarization feed of a dual-pol feed antenna and Port 2 connected to the vertical polarization feed as shown in Figure 43. With this configuration the antenna is simultaneously used in both receive and transmit mode. For the remainder of this chapter, the S-parameter designations are transformed from the port-centric designation to the corresponding co and cross polarization designations: $11 \rightarrow hh$, $12 \rightarrow hv$, $21 \rightarrow vh$, and $22 \rightarrow vv$.

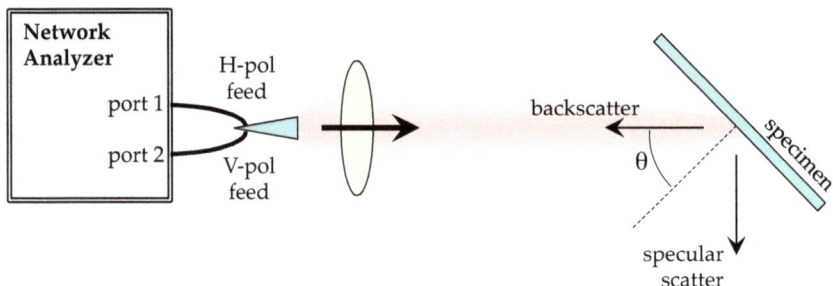

Figure 43   Configuration for cross-polarization measurements.

A dihedral corner reflector can be made from two metal plates at right angles to each other in a way that incoming energy is reflected directly (or specularly) back to a transmitting antenna. For the focused beam system, if the dihedral is built large enough to capture the entire quasi-plane wave focused onto it, all the energy is expected to be reflected back as an identical quasi-plane wave. When the junction of the plates is oriented vertically, a right angle dihedral has a reflection characteristic such that a horizontally polarized incident wave experiences a voltage reflection coefficient of $\Gamma = -1$, while a vertically polarized wave sees a reflection coefficient of $\Gamma = 1$. The opposite signs of the reflection coefficient are due to a 180° phase shift difference (or rotation) from the "double bounce" a wave encounters during the round trip inside the dihedral depending upon its polarization. This characteristic provides a polarization dependent reflection coefficient whereas the reflection coefficient of a flat metal plate is independent of polarization.

Because of the polarization dependent reflection coefficient, rotation of the dihedral about its center axis provides a strong cross-polarization reference signal, useful for a full cross-polarization calibration algorithm. In particular, the electric field of the signal incident upon a vertically oriented dihedral is rotated 180° for horizontal polarization, and is not rotated for vertical polarization. Thus the scattering of a rotated dihedral is determined by separating the incident signal into both its horizontal and vertical components, then the horizontal component is rotated 180° and recombined to obtain the total received signal. The vector components of the received signal are analyzed with respect to the cross- and co-polarization axes.

Measurement of complex voltage by network analyzers allows a representation of the scattering from complex targets as a polarization scattering matrix (PSM),

$$\begin{bmatrix} a^- \\ b^+ \end{bmatrix} = \begin{bmatrix} S_{hh} & S_{hv} \\ S_{vh} & S_{vv} \end{bmatrix} \begin{bmatrix} a^+ \\ b^- \end{bmatrix} \qquad (6.32)$$

where the a wave corresponds to horizontal polarization, and the b wave corresponds to vertical polarization. The $+$ and $-$ designations are defined in Figure 44, along with the geometry of the dihedral and the orientation (ϕ) of its axis relative to the measurement system. The PSM at each frequency is specified by eight scalar quantities, four amplitudes, and four phases where one phase angle is arbitrary and is used as a reference for the other three.

Based on concepts originally for remote sensing, Chen-Chu-Chen [65] developed a calibration technique that uses three calibration standards in an anechoic chamber to determine the PSM of an unknown target. The relationship between the actual target PSM and the measured PSM can be described as $\mathbf{S}^m = \mathbf{X} + \mathbf{RST}$ where $\mathbf{X}$ is an isolation error matrix resulting from residual reflections and coupling between transmitting and receiving channels when no target is present. $\mathbf{T}$ and $\mathbf{R}$ are transfer matrices that account for frequency response, mismatches, and cross-polarization coupling, $\mathbf{S}$ is the target scattering matrix and $\mathbf{S}^m$ is the measurement. Barnes introduced this RST model in 1986 [66] and in expanded form the measured signal is,

$$\begin{bmatrix} S_{hh}^m & S_{hv}^m \\ S_{vh}^m & S_{vv}^m \end{bmatrix} = \begin{bmatrix} X_{hh} & X_{hv} \\ X_{vh} & X_{vv} \end{bmatrix} + \begin{bmatrix} R_{hh} & R_{hv} \\ R_{vh} & R_{vv} \end{bmatrix} \begin{bmatrix} S_{hh} & S_{hv} \\ S_{vh} & S_{vv} \end{bmatrix} \begin{bmatrix} T_{hh} & T_{hv} \\ T_{vh} & T_{vv} \end{bmatrix} \tag{6.33}$$

which can be rearranged when solving for a measured target to

$$S = R^{-1}(S^m - X)T^{-1} \tag{6.34}$$

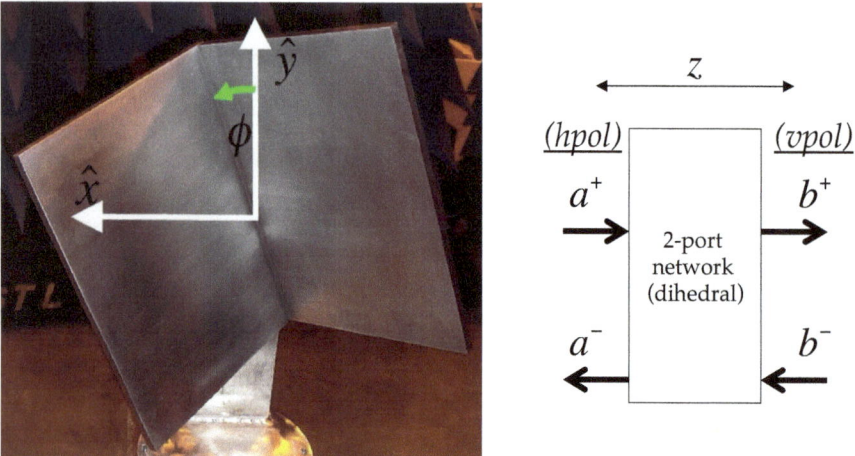

Figure 44   **Geometry of dihedral calibration standard (left) and two-port network definitions for a PSM (right).**

This equation represents a set of coupled nonlinear equations. Thus calibration involves finding X, R, and T. X is simply the isolation measurement (matched load) where $S = 0$. Yueh proposed a solution that obtains normalized quantities of the R and T matrices [67]. The solution requires three different calibration standards: flat metal plate, dihedral, and rotated dihedral. The PSM of a flat metal plate is

$$S_1 = \begin{bmatrix} -1 & 0 \\ 0 & -1 \end{bmatrix} \tag{6.35}$$

Inversion of this calibration standard is possible because the diagonal elements of the flat plate PSM are non-zero. The metal plate standard establishes a relation between R and T of $R = S_1^m T^{-1}$, so therefore derivation of T will also give R. The normalized T matrix can be defined as,

$$\mathbf{T} = \begin{bmatrix} T_{hh} & T_{hv} \\ T_{vh} & T_{vv} \end{bmatrix} = T_{vv} \begin{bmatrix} w & w/u \\ v & 1 \end{bmatrix} = T_{vv} \mathbf{T}' \qquad (6.36)$$

Solutions of u and v are obtained using the measurement of the second independent calibration standard, with the requirement that the diagonal terms of the standard are independent of each other. The dihedral corner reflector fills this requirement with a PSM of

$$\mathbf{S}_2 = \begin{bmatrix} -1 & 0 \\ 0 & 1 \end{bmatrix} \qquad (6.37)$$

where one polarization acts as an electrical equivalent of a short, and the orthogonal polarization acts as an electrical open. A solution for w is obtained using the third calibration standard, the rotated $(\phi = 22.5°)$ dihedral. This rotated dihedral provides independent cross-polarization information on the diagonals necessary for solving w. The PSM of a rotated dihedral is

$$\mathbf{S}_3 = \begin{bmatrix} -\cos 2\phi & \sin 2\phi \\ \sin 2\phi & \cos 2\phi \end{bmatrix} \qquad (6.38)$$

$\mathbf{R}$ and $\mathbf{T}$ are then solved by evaluating Equation 6.34 with these PSM expressions of the various calibration standards (noting that T_{vv} is cancelled out, and the result is normalized to the S_{vv} component of the flat plate calibration).

7. ADVANCED METHODS

APERTURES

As described in previous chapters, accurate material characterization can be achieved by free-space illumination when focusing lenses confine the incident energy to within the interior portion of a sample. However, there are often constraints on the maximum size of a material specimen such that there may be significant over-illumination of the specimen. This is particularly true at lower frequencies, where the large wavelength drives ever-larger illumination areas. In some cases, it might be possible to overcome this limitation by encompassing a too-small specimen, within a larger conductive ground plane. This concept of positioning an aperture around a specimen is illustrated notionally in Figure 45. It is similar to the older admittance or transmission tunnel methodology, in which a specimen along with the transmit and receive antenna are contained within an absorber-lined box. Instead with the apertured focused beam, the ground plane that includes the aperture in the figure is sized so that it is significantly bigger than the illuminating beam diameter, eliminating the need for surrounding absorber. Measurement with an aperture such as shown in Figure 45, represents a compromise between accuracy and small specimen size.

To illustrate the effects of apertures, Figure 46 shows the transmission coefficient of a thin resistive specimen that would be measured by a focused beam system. These data were calculated by a full wave finite difference time domain (FDTD) solver. The focused beam was simulated in this code by adding a series of weighted plane waves with a

Gaussian plane wave distribution. The transmission coefficient was determined by propagating the focused beam through the specimen sheet. In the simulations presented here, a Gaussian beam with a beam waist (radius) of $0.8\lambda_\omega$ was used; which approximates the beam from a 1.8 m (72 inch) diameter lens apparatus at the Georgia Institute of Technology. The transmission coefficient data were computed as a ratio to the transmission of a clear-site (no sample) to minimize grid dispersion errors.

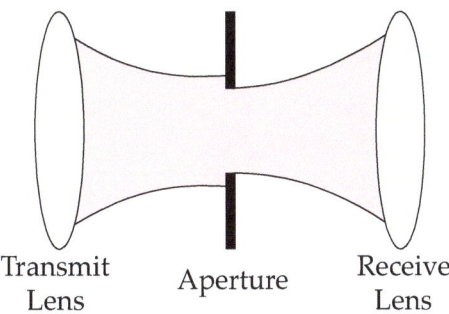

Figure 45   Notional sketch of aperture in focused beam system.

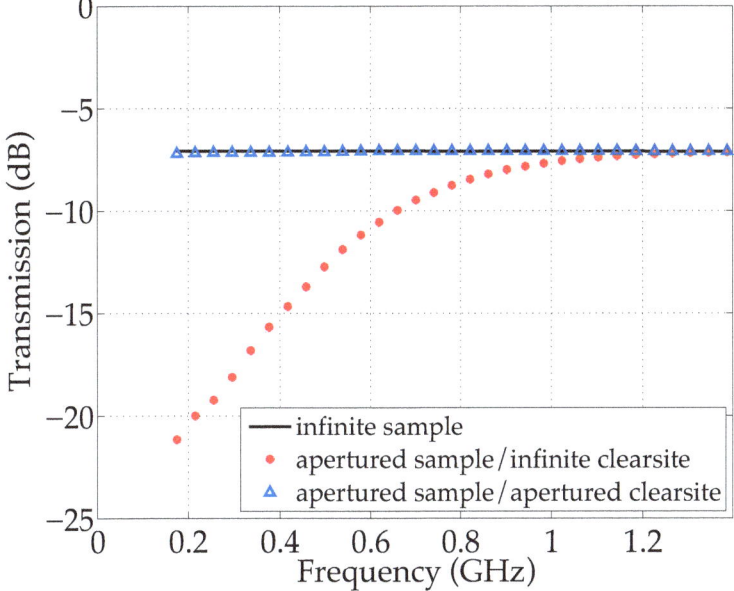

Figure 46   FDTD calculated transmission coefficient of a 150 Ω/square sheet illuminated by a Gaussian beam.

At low frequencies, the sample height and width may be smaller than the incident beam. Thus the measurement methodology simulated in Figure 46 included a metal ground plane with a rectangular aperture to limit propagation around the sample, as pictured in Figure 45. The calculated amplitude transmission coefficients for a 150 Ω/square sheet are shown in Figure 46. Data are shown for an infinite sheet with no aperture (solid line), an apertured sheet calibrated to an infinite clear-site, and an apertured sheet calibrated to an apertured clear-site. The aperture was 1.1 x 1.2 meters and the data were processed with a 4 ns time-domain gate to reduce aperture-ringing effects. These data show that the beam begins over-illuminating the sample below one GHz, but that calibrating to an apertured clear-site is an effective way to correct for this over-illumination.

When properly calibrated, the apertured transmission amplitude shows negligible errors from over-illumination, however significant phase variations occur from the aperture edge scatter. These errors are more intuitively viewed in terms of effective measured impedance. The effective sheet impedance, Z is determined from the transmission coefficient, S_{21}, by,

$$Z = \left(\frac{Z_0}{2}\right)\frac{S_{21}}{1 - S_{21}} \qquad (7.1)$$

where Z_0 is the impedance of free space (377 ohms). The calculated impedances from an apertured 150 Ω/square sheet calibrated to an apertured clear-site are shown in Figure 47. Two cases of apertured specimen are shown: no gap, where the impedance sheet is electrically connected to the ground plane; and a 2 cm gap, where there is an air gap between the periphery of the sheet and the edge of the aperture. When there is good electrical contact between the sample and the ground plane, the real impedance is very close to the actual impedance of the sheet, but a small inductance (positive imaginary impedance) occurs at the lowest frequencies. With an air-gap between the sample edges and the ground plane, the real impedance increases at lower frequencies and the imaginary impedance shows a significant capacitance (negative imaginary impedance) from interaction between the sample and ground plane edges across the gap.

The inductive and capacitive effects caused by the aperture edges are confirmed by the model-measurement comparison of Figure 48. The FDTD data in these figures were calculated for a 440 Ω/square sheet. The measured sample was a graphite filled polyimide sheet (DuPont). Both measured and simulated data were processed with a 4 ns time-domain gate. Both sets of data show good qualitative agreement. However, there are some quantitative differences due to frequency dispersion in the actual sample impedance (The FDTD model assumed an impedance that was the same for all frequencies). In addition, the beam was assumed to be a symmetrical Gaussian beam with a constant $k_0 w_0$ in the FDTD simulations, however the actual measurement system has a slightly elliptical beam with some frequency dispersion in $k_0 w_0$.

These results show that with appropriate calibration, the aperture is effective at extending the focused beam methodology to samples that are roughly a wavelength across. Of course, there is still a need to go even further and measure ever-smaller specimens, which adds even more error from over-illumination. One approach is to fully fill the larger metallic bound aperture with strips of both the unknown material along with previously measured known materials. An example geometry is shown in Figure 49, which shows a metallic aperture fixture with a small strip of the unknown material, Z_2 surrounded by strips of a known impedance material, Z_1. This geometry would be useful if laboratory samples could be obtained only in narrow strips.

The transmission coefficient from this three-strip fixture will result in an effective impedance that is an average of the two known and one unknown strips. It can be modeled as a weighted area average of the voltage transmission coefficients of each impedance region,

$$T_{effective} = W_{Z_2} T_2 + \left(1 - W_{Z_2}\right) T_1 \tag{7.2}$$

where T_1 is the voltage transmission coefficient of the known impedance strips and T_2 is for the unknown. The weight for the middle (unknown) impedance strip is proportional to the voltage across the strip, which is calculated by integrating the Gaussian beam taper over the middle strip width $(2r)$,

$$W_{Z_2} = \frac{\int_0^r e^{-x^2/w_0^2} dx}{\int_0^\infty e^{-x^2/w_0^2} dx} = \frac{w_0 \sqrt{\pi/4}\, \mathrm{erf}(r/w_0)}{w_0 \sqrt{\pi/4}} = \mathrm{erf}\left(\frac{r}{w_0}\right) \tag{7.3}$$

where w_0 is the beam waist of the Gaussian beam. Because the aperture is smaller than the total beam, this weight is also normalized to the total field going through the aperture. After including an empirical multiplicative factor, f_1 to account for diffraction and aperture shape effects, the resulting weight is,

$$W_{Z_2} = \frac{\mathrm{erf}(f_1 r/w_0)}{\mathrm{erf}(f_1 R/w_0)} \tag{7.4}$$

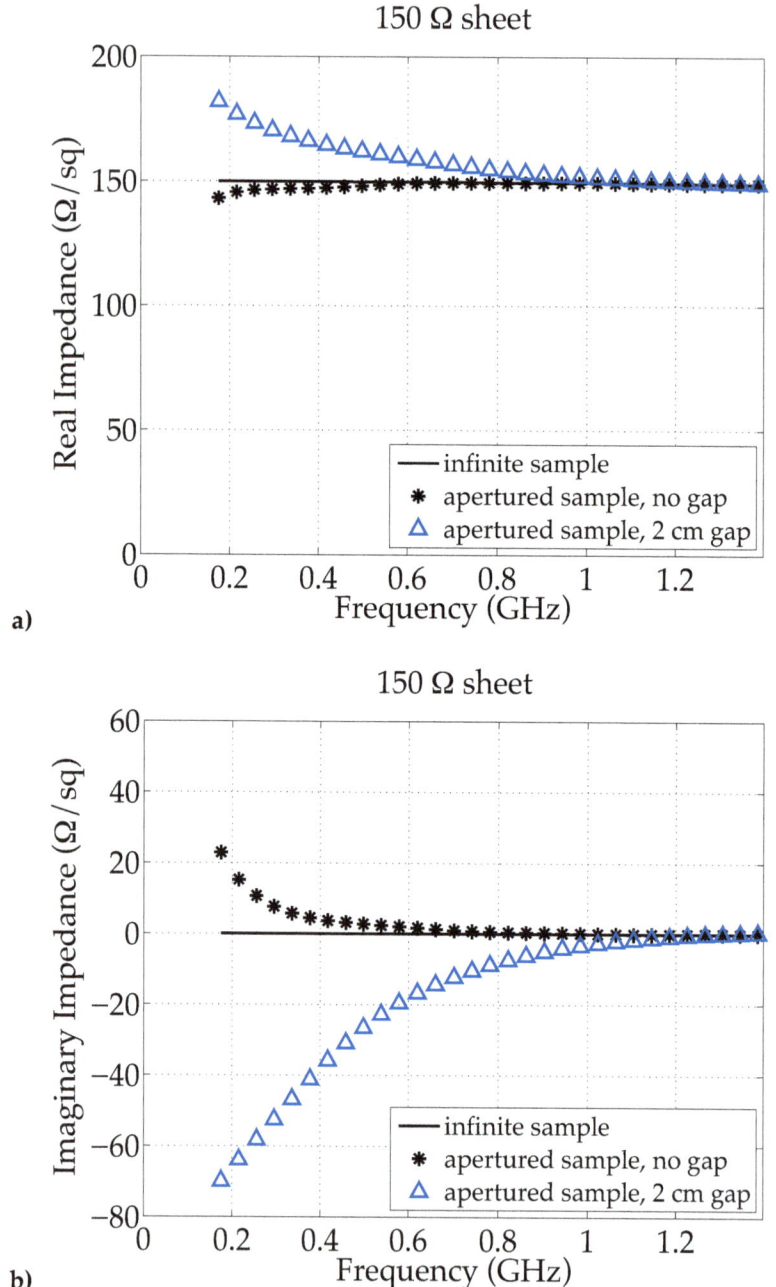

a)

b)

Figure 47   Effect of gap on a) real and b) imaginary sheet impedance
calculated for 150 Ω/square sheets in an aperture.

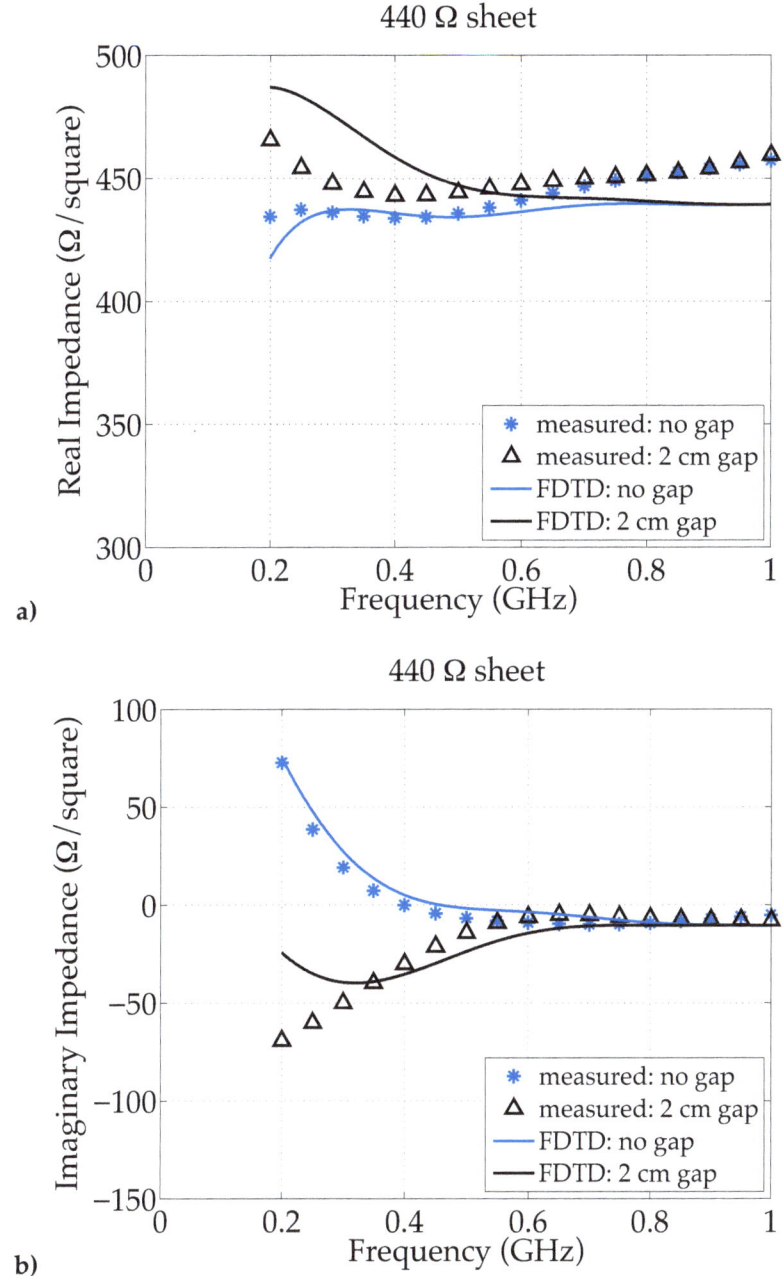

Figure 48   **Experimentally measured and FDTD calculated impedance with and without 2 cm gaps.**

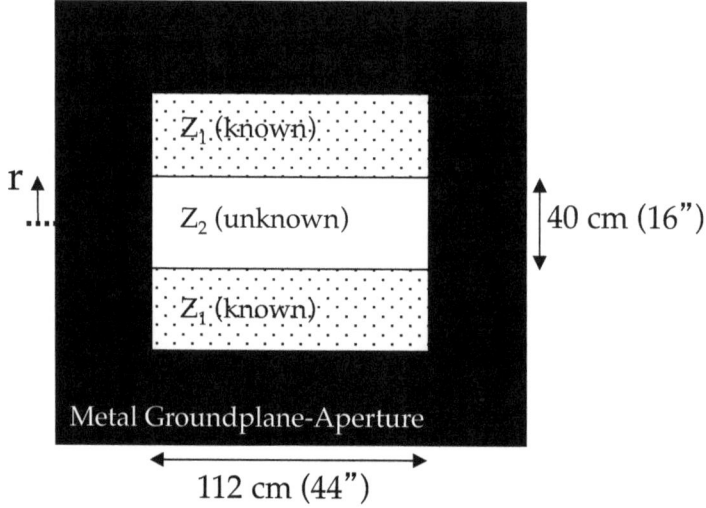

**Figure 49   Geometry for measuring impedance of a narrow strip
sample, Z_2. The rest of the aperture is filled with a known
impedance sheet, Z_1.**

Figure 50 shows that this weighted average model agrees with
FDTD calculations of effective impedance. These calculations were for an
assumed aperture size of 1.1 x 1.2 meters, where the width of each strip
was 0.4 meters. The two known strips had sheet impedances of 150 Ω/sq.
The FDTD calculations and semi-empirical model in Figure 50 show
agreement even when Z_1 and Z_2 differ by a factor of two. When fitting
measured data to FDTD simulations, the semi-empirical model can be used
to supplement the simulations thereby reducing the computational effort
needed.

NEAR-FIELD PROBE

Far-field scattering measurements can provide information on
radiated fields from an object, but supply little information on evanescent
fields such as surface waves or local cavity modes. However, these non-
radiating fields can contribute to radiated fields if local perturbations
appear, such as geometric discontinuities or electromagnetic property
variations. Near-field measurements of scattering bodies provide

additional insight into these scattering mechanisms by measuring both radiated and non-radiated fields. Near-field measurements can also detect covered or buried inhomogeneities in dielectric or magnetic media, and are used to measure evanescent fields in antennas and microwave circuits, providing information about their electromagnetic behavior. In addition to a near-field probe, such measurements also require a microwave source to excite the object under investigation. In some cases, a focused beam is beneficial since it can provide a controlled and localized illumination of an object or a portion of an object. This section describes a method for using a focused beam system combined with a probe to investigate microwave scatter phenomena associated with linear discontinuities such as edges.

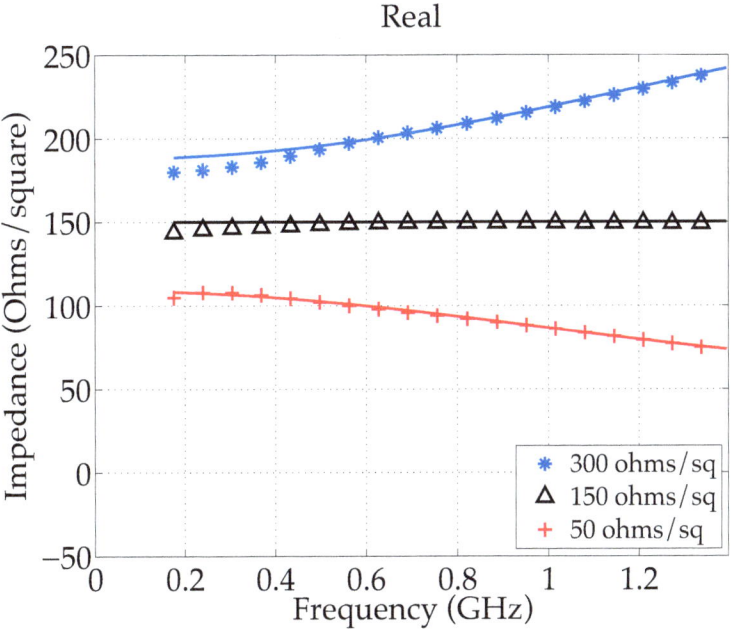

Figure 50   Effective impedance measured with aperture fixture of Figure 49 for different Z_i values. Symbols are FDTD calculations and lines are semi-empirical model.

When investigating scatter, it is useful to assume that the illuminating energy can be approximated as a plane wave. Additionally, accurate measurement of the scatter from a feature or inhomogeneity requires that the extent of the illuminating beam be greater than the physical dimension of the scattering feature. For example, a rounded edge with radius of curvature r, must be encompassed within the illuminating beam (including both the width and depth). This is analogous to compact range RCS measurements where the target under test must be contained within the 'quiet zone'.

A measure of the depth of focus is the Rayleigh length (or Rayleigh range), defined previously as the distance from the beam waist location to where the spot size has increased to $w(Z_R) = \sqrt{2}w_0$. It is given by,

$$Z_R = \frac{k}{2}w_0^2 \qquad\qquad (7.5)$$

For a focused beam where $w_0 \sim 2\lambda_0$, and assuming the physical size of the scattering feature must be less than $2Z_R$, then $2Z_R < 8\pi\lambda_0$, or approximately 25 wavelengths. A more restrictive dimension is the width of the illuminating beam. In analogy with the compact range "quite zone", we can define a similar quiet zone within the focused beam where the amplitude taper decreases by no more than 3 dB. Note that even if these rules of thumb are followed, there may still be finite size effects [68], so data interpretation should always consider the presence of these effects.

The rest of this section illustrates near field probe measurements combined with focused beam illumination by describing measurements of the near-fields around a simple, linear edge. In the following data, a focused beam illuminated the edge of an aluminum sheet with dimensions, 122 cm wide x 91 cm tall x 0.041 cm thick. The edge is positioned at the beam's focus in the region of approximately constant phase. A small loop probe is spatially scanned either linearly or in a raster pattern in the near-field region immediately in front of the sample with a separation of approximately 4 mm between the sample surface and the probe center. Thus the probe is $\leq \lambda/9$ from the sample surface within the measured frequency range. The induced signal in the probe is proportional to the total local H-field, which is a dependent on spatial position and frequency. Note that this assumes that the probe does not significantly shadow the incident beam onto the specimen under test. This is helped primarily by orienting the feed cables to the probe to be orthogonal to the incident E-field polarization, thus minimizing probe scatter.

In the measurement procedure, a scan is made with the specimen in place, and a second scan is made with no specimen. The second, or 'clear-sight' scan is proportional to the incident field while the first scan includes both incident and scattered fields. The measured field data is normalized to the incident (clear sight) field at the center of the beam and the data is expressed as normalized near field amplitude or phase. The experimental geometry for edge illumination of a simple sheet is shown in Figure 51. For the data shown, the probe was scanned linearly across the edge and parallel to the plane of the sheet.

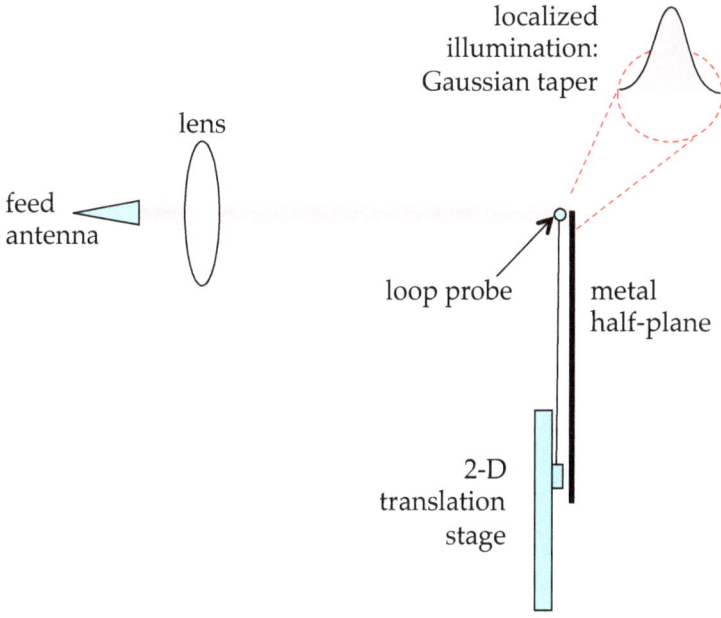

Figure 51   Geometry of focused beam and near field probe measurements.

The total field measured in front of a metal half plane with a normal incidence, H-beam is shown in Figure 52. The data are shown in terms of H-field, normalized to the field of the center of the incident beam as a function of position and frequency. The edge of the metal sheet is at position = 0 and extends to position > 0. Thus the higher field amplitude apparent at positions > 0, is due to the sum of the incident and specular reflected energy, while at positions < 0, there is no metal sheet and therefore no specular reflection.

Both the total field and incident field are measured with the probe by scanning both with and without the half sheet present. Vector subtracting the incident field from the total field results in the scattered field, which is shown in Figure 53. As before the data are normalized to the peak incident H-field (clear site). The scattered field shows that there is little energy at positions < 0, which are away from the metal sheet. Furthermore, an interference pattern is evident, due to the superposition of an edge excited wave and the specular wave at positions > 0.

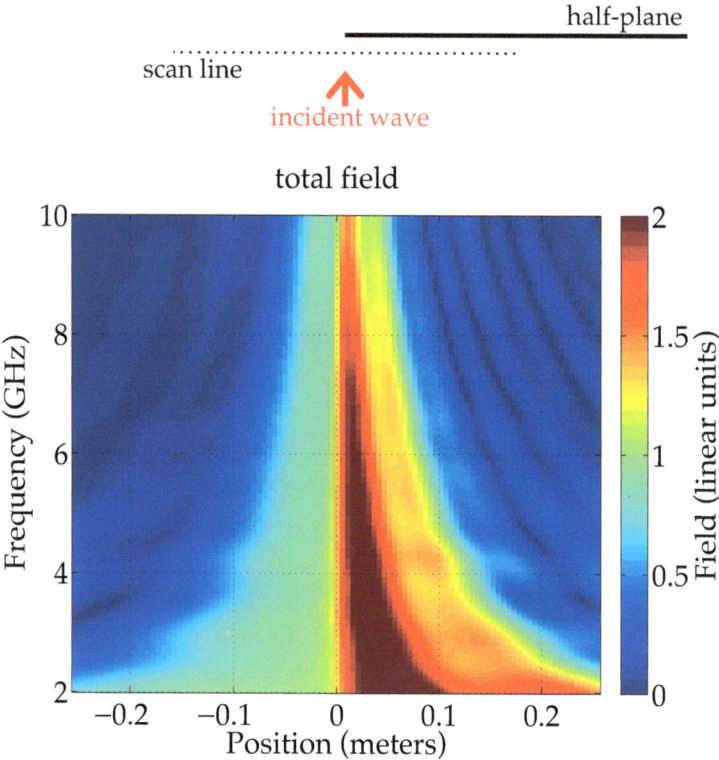

Figure 52   **Amplitude of total H-field (incident + scattered) from half plane illuminated at normal incidence, TM (H-beam) polarization, normalized to incident field at center of beam.**

An alternative way to view the measured probe data is in a "ω-β" diagram of the plane wave spectrum. Since data are acquired as a function of position on the plate, the plane wave spectrum can be calculated via Fourier transform with respect to spatial position [61]

$$\frac{F_y(k_x)}{F_0} = \int_{-\infty}^{\infty} H_x(x)e^{ik_x x}dx \qquad (7.6)$$

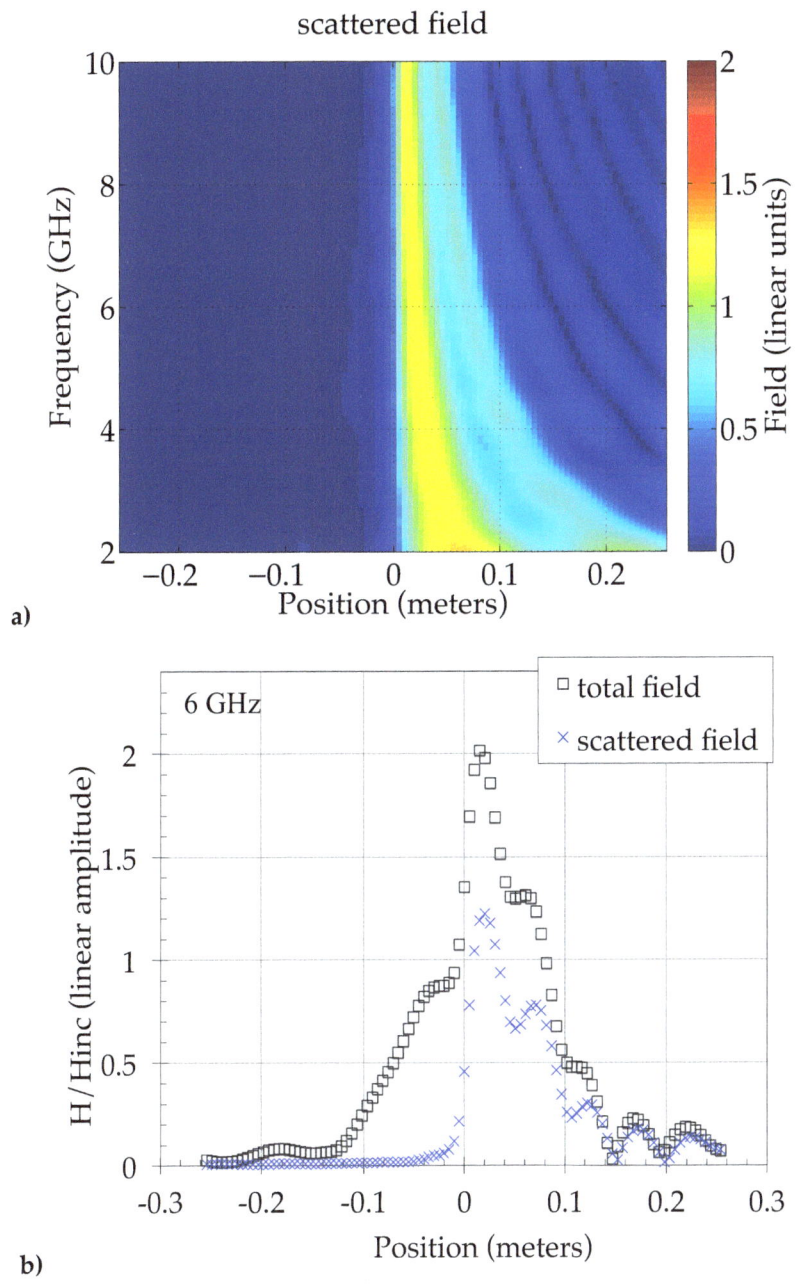

Figure 53   a) Amplitude of scattered H-field from half plane
illuminated at normal incidence, TM (H-beam)
polarization, normalized to incident field at center of
beam. b) comparison of total and scattered field at 6 GHz.

Figure 54 shows the plane wave spectra of the scattered H-fields for H-beam illumination of the edge at normal incidence. This is the polarization where the incident electric field is perpendicular to the edge. $k_x = \pm k_0$ 'light lines' are indicated as thin black lines on the graph. Thus plane wave energy outside of these lines constitute evanescent modes that do not propagate to the far field.

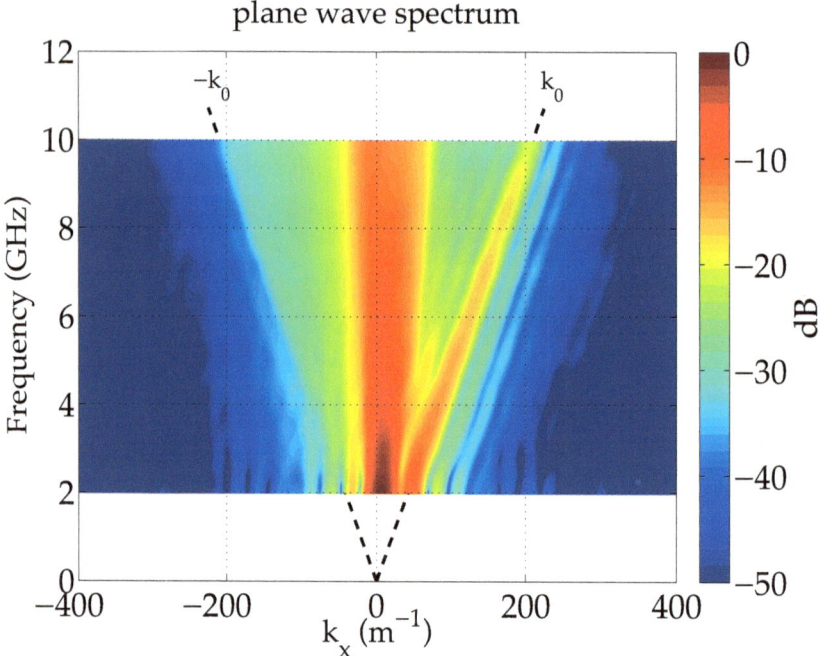

Figure 54   **Plane wave spectra of scattered H-fields from half plane edge, illuminated at normal incidence, H-beam (TM). Colors indicate dB amplitude.**

The data in Figure 54 indicate both specular reflection from the metal sheet and edge wave scatter. Because of the normal incidence, the spectra show a primary lobe centered at $k_x = 0$. The width of this k-vector lobe depends on the width of the illuminating beam and reflects the plane wave distribution of the incident beam. Superimposed with this specular lobe is a broader distribution of plane waves that span the k-space between the light lines at approximately 20 to 30 dB below the specular lobe. These broader spectra are due to diffracted energy from the edge. In these data the strongest component of diffracted propagation is in the $k_x = k_0$ direction, which is the direction along the surface of the metal plane (i.e. a surface travelling wave).

These data illustrate the power of combining a near-field probe measurement with a focused beam system. In this case, we have a method for experimentally characterizing the physical phenomena responsible for scatter behavior from edge discontinuities. Similarly this method can be used to understand the electromagnetic behavior for applications ranging from component scatter to frequency selective surfaces and radomes.

REFERENCES

1. A. Sihvola, Electromagnetic mixing formulas and applications, IEE, London, UK, (1999)
2. P. Debye, Polar Molecules, Dover, New York, (1929)
3. P. Drude, The Theory of Optics, Dover, New York, Translated by C. Riborg Man & R.A. Millikan, (1902)
4. K.S. Cole, R.H. Cole, "Dispersion and Absorption in Dielectrics," J. Chem. Phys., 9, 341-351, (1941)
5. S. Havriliak Jr., S.J. Havriliak, "Unbiased Modeling of Dielectric Dispersions," in Dielectric Spectroscopy of Polymeric Materials, Chapter 6, Ed. J.P. Runt, J.J. Fitzgerald, 175-200, (1997)
6. R.M. Fuoss, J.G Kirkwood, "Electrical Properties of Solids. VIII. Dipole Moments in Polyvinyl Chloride-Diphenyl Systems," J. Am. Chem. Soc., 63, 385-401, (1941)
7. D.W. Davidson, R.H. Cole, "Dielectric Relaxation in Glycerine," J. Chem. Phys., 18, 1417, (1950)
8. F. Kohlrausch, Pogg. Ann. Phys. Chem., 119, 337, (1963)
9. W. Williams, D.G. Watts, "Non-Symmetrical Dielectric Relaxation Behavior Arising from a Simple Empirical Decay Function," Trans. Faraday Soc., 66, 80-85, (1970)
10. A.K. Jonscher, Dielectric Relaxation in Solids, Chelsea Dielectric: London, (1983)
11. D.T. Emerson, "Jagadis Chandra Bose: Millimetre Wave Research in the Nineteenth Century," IEEE Trans MTT, 45(12), 2267-2273, (1997)
12. R.M. Redheffer, "The Measurement of Dielectric Constants," in Technique of Microwave Measurements, C.G. Montgomery ed., McGraw-Hill, New York, 561-678, (1947)
13. R.M. Redheffer, "Microwave Antennas and Dielectric Surfaces," J. Appl. Phys., vol 20, 397-411, April (1949)
14. T.E. Talpey, "Optical Methods for the Measurement of Complex Dielectric and Magnetic Constants at Centimeter and Millimeter Wavelengths," L'Onde Electrique, October (1953)

15. R.E. Hiatt, E.F. Knott, T.B. Senior, "A Study of VHF Absorbers and Anechoic Rooms," University of Michigan Report 5391-1-F for NASA contract NASr-54(L-1), Langley Research Center, Hampton, VA, February (1963)
16. NRL Public Affairs, "75 Years Naval Research Laboratory," press release, (1998)
17. D.C. Stallings, "Resistive Sheet Measurements," Report for Project A-2583, Georgia Institute of Technology, (1980)
18. F.E. Boyd, "Converging Lens Dielectric Antennas," NRL Report 3780, Naval Research Laboratory, Washington DC, DTIC ADB801103, (1950)
19. R.I. Primich, "Microwave Techniques for Hypersonic Ballistic Ranges," Planetary and Space Science, v. 6, 186-195, 1961
20. R.I. Primich, F.H. Northover, "Use of Focused Antenna for Ionized Trail Measurements: Part 1. Power Transfer Between Two Focused Antennas," IEEE AP, 112-118, March 1963
21. W.M. Braezeale, "Method and Apparatus for Measuring Moisture Content", U.S. Patent # 2,659,860, filed Aug 27, 1949, awarded Nov. 17, 1953
22. W. Culshaw, "A Spectrometer for Millimetre Wavelengths," Proc. IEE – Part IIA: Insulating Materials, 100(3), 1953
23. G. Goubau, "Beam-Waveguide Antenna," U.S. Patent 2,994,873, (1961)
24. J Datlov, J. Musil, F. Zacek, "Beam Width of Two Antenna Systems for Plasma Diagnostics," Czechoslovak Journal of Physics, 15(10), 766-768, (1965)
25. H.L Bassett, "A Free-Space Focused Microwave System to Determine the Complex Permittivity of Materials o Temperatures Exceeding 2000 C," Ref. Sci. Instr., 42(2), 200-204, 1971
26. J.L Pentecost, "Electrical Evaluation of Radome Materials," in Radome Engineering Handbook, Design and Principles, J.D. Walton ed., Marcel Dekker, New York, 1970
27. J.D. Walton, S.H. Bomar, H.L. Bassett, "Evaluation of Materials in a High Heat Flux Radiant Thermal Energy Environment," AMMRC CTR 73-16, Final Report for Contract DAAG46-72-C-0189, 1973
28. J. Musil, F. Zacek, A. Burger, J. Karlovsky, "New Microwave System to Determine the Complex Permittivity of Small Dielectric and Semiconducting Samples," 4th European Microwave Conference, 66-70, 1974
29. Z. Zhang, "Design of the Broadband Admittance Tunnel for High Fidelity Material Characterization," Ph.D. dissertation, Arizona State University, 2005
30. P.F. Goldsmith, Quasioptical Systems, Gaussian Beam Quasioptical Propagation and Applications, IEEE Press, (1998)
31. P.F. Goldsmith, T. Itoh, K.D. Stephan, "Quasi-Optical Techniques," in Handbook of Microwave and Optical Components, Volume 1, K. Chang ed., Wiley, 344-363, (1989)
32. J. Musil, F. Zacek, Microwave Measurement of Complex Permittivity by Free Space Methods and Their Applications, Elsevier, (1986)

33. D. Marcuse, <u>Light Transmission Optics</u>, 2ᵈ Ed., Van Nostrand, (1982)
34. J. Peatross, M. Ware, <u>Physics of Light and Optics</u>, Brigham Young University, (2009)
35. I. Bruce, "ABCD transfer matrices and paraxial ray tracing for elliptic and hyperbolic lenses and mirrors," European J. Phys., 27, 393-406, 2006
36. S. Gangopadhyay, S Sarkar, "ABCD matrix for reflection and refraction of Gaussian light beams at surfaces of hyperboloid of revolution and efficiency computation for laser diode to single-mode fiber coupling by way of a hyperbolic lens on the fiber tip," Applied Optics, 36(33), 8582-8586, (1997)
37. D.K. Ghodgaonkar, V.V. Varadan, V.K. Varadan, "Free-Space Method for Measurement of Dielectric Constants and Loss Tangents at Microwave Frequencies," IEEE Trans. I&M, 37(3), 789-793, (1989)
38. Harris, Fredric J., "On the Use of Windows for Harmonic Analysis with the Discrete Fourier Transform," Proc. IEEE, 66(1), 51-83 (1978).
39. M. Born, WE. Wolf, <u>Principles of Optics,</u> 6ᵗʰ Ed. Cambridge Univ. Press, (1980)
40. A.M. Nicolson, G. Ross, "Measurement of Intrinsic Properties of Materials by Time Domain Techniques," IEEE Trans. Instrum. Meas., 19, 377-382, (1970)
41. W.B. Weir, "Automatic Measurement of Complex Dielectric Constant and Permeability at Microwave Frequencies," Proc. IEEE, 62, 33-36, (1974)
42. J. Baker-Jarvis, "Transmission/Reflection and Short-Circuit Line Permittivity Measurements," NIST Technical Note 1341, (1990)
43. A.-H. Boughriet, C. Legrand, A. Chapoton, "Noniterative Stable Transmission/Reflection Method for Low-Loss Material Complex Permittivity Determination," IEEE Trans. Microwave Theory and Techniques, 45(1), 52-57, (1997)
44. W.H. Press, S.A. Teukolsky, W.T. Vetterling, B.P. Flannery, <u>Numerical Recipes 3ᵈ Edition: The Art of Scientific Computing</u>, Cambridge University Press, (2007)
45. J. Baker-Jarvis, M.D. Janezic, J.H. Grosvenor, Jr., R.G. Geyer, "Transmission/Reflection and Short-Circuit Line Methods for Measuring Permittivity and Permeability," NIST Technical Note 1355, (1992)
46. S. Roberts, A. Von Hippel, "A New Method for Measuring Dielectric Constant and Loss in the Range of Centimeter Waves," J. Appl. Phys. **17**, 610-616 (1946)
47. E.F. Knott, J.F. Shaeffer, M.T. Tuley, <u>Radar Cross Section</u>, 2ᵈ ed., Artech House, 1993
48. S. Ramo, J.R. Whinnery, T. Van Duzer, <u>Fields and Waves in Communication Electronics</u>, 3ᵈ ed., Wiley, (1994)
49. D. Xiu, "Fast Numerical Methods for Stochastic Computations: A Review," Comm. In Computational Phys., 5(2-4), 242-272, (2009)

50. J.R. Taylor, <u>An Introduction to Error Analysis</u>, Oxford Univ. Press, (1982)
51. L.E.R. Petersson, G.S. Smith, "An Estimate of the Error Caused by the Plane-Wave Approximation in Free-Space Dielectric Measurement Systems," IEEE Trans. AP, 50(6), 878-887, (2002)
52. L.E.R. Petersson, "Analysis of Two Problems Related to a Focused Beam Measurement System," PhD Dissertation, Georgia Institute of Technology, November 2002
53. D.K. Rytting, "Network Analyzer Accuracy Overview," ARFTG Conference Digest – Fall, 58th, 40, 1-13, (2001)
54. K. Wong, "Network Analyzer Calibrations – Yesterday, Today and Tomorrow," Symposium Digest, IEEE MTT-S International, 19-25, (2008)
55. J.D. Jackson, <u>Classical Electrodynamics</u>, 3rd ed., Wiley, (1998)
56. <u>Introduction to Complex Mediums for Optics and Electromagnetics</u>, W.S. Weiglhofer & A. Lakhtakia eds., SPIE Press, (2003)
57. B.A. Munk, <u>Frequency Selective Surfaces</u>, Wiley, (2000)
58. J.W. Schultz, E.J. Hopkins, J.G. Maloney, E.J. Kuster, M.P. Kesler, "A Focused-Beam Methodology for Measuring Microwave Backscatter," Microwave and Optical Tech. Lett., 42(3), 201-205, 2004
59. R.E. Collin, "Scattering of an incident Gaussian beam by a perfectly conducting rough surface," IEEE Trans. Antennas and Propagation, vol. 42, (1994), 70-74
60. M.W. Long, <u>Radar Reflectivity of Land and Sea</u>, third ed., Artech House, 2001
61. G.S. Smith, <u>An Introduction to Classical Electromagnetic Radiation</u>, Cambridge Univ. Press, 1997
62. D.M. Kerns, <u>Plane-Wave Scattering-Matrix Theory of Antenna-Antenna Interactions</u>, National Bureau of Standards, Monograph 162, U.S. Gov Printing Office, 1981
63. G.T. Ruck, D.E. Barrick, W.D. Stuart, C.K. Krichbaum, <u>Radar Cross Section Handbook</u>, Volume 1, Pleneum Press, New York, (1970)
64. Larson, T., "A Survey of the Theory of Wire Grids", IRE Trans. MTT, 191-201, May 1962
65. Chen, T., Chu, T., and Chen, F., "A new calibration algorithm of wide-band polarimetric measurement system", IEEE Trans. Antennas and Propagation, 1991, August, pp. 1188-1192
66. R. M. Barnes, "Antenna polarization calibration using in-scene reflectors," Proceedings of the Tenth DARPA/Tri-Service Millimeter Wave Symposium, U.S. Army Harry Diamond Lab., Adelphi, MD, April 8-10, 1986
67. S. H. Yueh, J. A. Kong, and R. T. Shin, "Calibration of Polarimetric Radars Using In-Scene Reflectors," Progress in Electromagnetics Research (PIER 3 – Polarimetric Remote Sensing), J. A. Kong Editor, Elsevier Publishing, New York, 1990
68. L.E.R. Petersson, G.S. Smith, "On the Use of a Gaussian Beam to Isolate the Edge Scattering From a Plate of Finite Size," IEEE Trans. AP, 52(2), 505-512, (2004)

GLOSSARY OF SYMBOLS

a	Semi-major axis of a hyperbola
A	Area
A, B, C, D	Parameters used in ray transfer matrix formalism
e	= 2.71828 ..., Eulers number
E	Electric field
f	Focal length or frequency (in Hertz) or iteratively solved function for inversion, depending on context
F	Plane wave spectrum
g	Iteratively solved function for inversion
i	Denotes imaginary number
I_{ij}	Network scattering parameter for isolation calibration standard
k	$= 2\pi\sqrt{\varepsilon_r \mu_r}/\lambda$, the wavenumber
k_0	$= 2\pi/\lambda$, the wavenumber in free space
l	length
M	Ray transfer matrix
n	$= \sqrt{\varepsilon_r \mu_r}$, the index of refraction

N	Number of points
P	Power
q	$= z + iZ_R$, the complex beam parameter
r	Radial coordinate
R	Radius of curvature or power reflection coefficient or scattering matrix, depending on context
R_{ij}	Network scattering parameter for response calibration standard; or R-matrix parameters, depending on context
S_{ij}	Network scattering parameter
t	Time or thickness, depending on context
t_i	Maximum thickness of a plano-convex lens
T	$= e^{-ikt}$, propagation factor of a wave through a medium or power transmission coefficient, depending on context
V	Voltage
w	The beam radius along a Gaussian beam
w_0	The beam waist of a Gaussian beam (the radius at the beam focus)
W	Power per unit area
x, y, z	Cartesian coordinates
X	Distance from origin to any position along surface of lens
Z_R	$= \frac{k}{2} w_0^2$, the Rayleigh range or confocal parameter
Z	Impedance. Sometimes this designates sheet impedance in ohms/square and sometimes a bulk impedance in ohms, depending on context.
α	Empirical fit parameter used in dielectric relaxation models
β	Empirical fit parameter used in dielectric relaxation models or propagation constant ($= 2\pi/\lambda$) depending on context
Γ	Fresnel reflection coefficient

$\tan \delta$	Dielectric loss tangent ($= \varepsilon''/\varepsilon'$)
δ	Damping parameter used in dielectric relaxation models or Dirac delta function depending on context
$\tan \delta_m$	Magnetic loss tangent ($= \mu''/\mu'$)
ε	Dielectric permittivity (note that sometimes this also designates the relative permittivity depending on context)
ε'	Real part of the dielectric permittivity ($\varepsilon = \varepsilon' - i\varepsilon''$)
ε''	Imaginary part of the dielectric permittivity ($\varepsilon = \varepsilon' - i\varepsilon''$)
ε_0	$= 8.854 \times 10^{-12}$ F/m, the dielectric permittivity of vacuum
$\varepsilon_r = \varepsilon/\varepsilon_0$	Relative dielectric permittivity (note that sometimes the subscript 'r' is dropped)
ε_R	Relaxed permittivity (in the high frequency limit) used in dielectric relaxation models
ε_U	Unrelaxed permittivity (in the low frequency limit) used in dielectric relaxation models
η	$= \sqrt{\mu/\varepsilon}$, the wave impedance
θ	Angle
λ	Wavelength
μ	Magnetic permeability (note that sometimes this also designates the relative permittivity depending on context)
μ'	Real part of the magnetic permeability ($\mu = \mu' - i\mu''$)
μ''	Imaginary part of the magnetic permeability ($\mu = \mu' - i\mu''$)
μ_0	$= 4\pi \times 10^{-7}$ H/m, the magnetic permeability of vacuum
$\mu_r = \mu/\mu_0$	Relative magnetic permeability (note that sometimes the subscript 'r' is dropped)
μ_R	Relaxed permeability (in the high frequency limit) used in magnetic relaxation models
μ_{int}	Intermediate permeability used in double-Lorentz magnetic relaxation model
μ_U	Unrelaxed permeability (in the low frequency limit) used in magnetic relaxation models

σ	$= \omega\varepsilon'' = \omega\varepsilon_0\varepsilon''_r$, electric conductivity or scattering cross-section (RCS) depending on context
τ	$= 1/\omega_0$, characteristic relaxation time in the Debye model; or Fresnel transmission coefficient, depending on context
ω	$= 2\pi f$, angular frequency in radians/second
ω_0	Characteristic relaxation frequency used in dielectric relaxation models
ω_p	Plasma frequency used in Drude model

INDEX